Data Acquisition and Analysis in Fisheries

NIPA® GENX ELECTRONIC RESOURCES & SOLUTIONS P. LTD.
New Delhi-110 034

About the Editors

Dr. T. Umamaheswari is an Assistant Professor specializing in Fisheries Economics within the Department of Fisheries Extension, Economics, and Statistics at the TNJFU-Fisheries College and Research Institute, Thoothukudi. With twelve years of experience in teaching and research, she has undertaken four research projects and developed instructional and practical manuals on Fisheries Economics, published 16 research papers, contributed to three book chapters, authored five conference proceedings, and written nine magazine articles. She has been honored with the Best Fisheries Scientist Award by Agro Environmental Development Society, Uttar Pradesh and Best Research Paper Award six times. In mentorship, she has guided four M.F.Sc. students and has held various administrative roles, including Students' Counsellor, Elected Representative in the Board of Studies (Fisheries Faculty), Technical Officer to the Dean, Secretary of the Thoothukudi Chapter of the Agricultural Scientific Tamil Society, New Delhi, Deputy Warden (Ladies Hostel), Students' Coordinator, Board of Studies Coordinator, and National Facilitator (MANAGE).

Dr. S. Balasundari, M.F.Sc., Ph.D has been holding the position of Dean, TNJFU- Dr. MGR Fisheries College and Research Institute, Thalainayeru since 2019. She has 31 years of professional experience including her service in the seafood industries, Agricultural Research Service, and State Agriculture University. Served as Head of the Department of Fish Processing Technology, Dr. MGR Fisheries College and Research Institute, Ponneri for 6 years. She has developed about 20 contemporary value-added products from finfish, shellfish, and seaweeds and commercialized 6 products. She published 9 books, 21 manuals, and 30 research articles. She has guided 10 research scholars in the field of Fish Processing Technology. She has received 11 awards conferred by university and scientific societies. She has completed six projects through state and national-level funding agencies and established 12 demonstration establishments as the major resource for various outreach programmes to benefit stakeholders.

Dr. G. Sugumar, former Vice-Chancellor of the Tamil Nadu Fisheries University, Nagapattinam started his career in teaching at the TNAU in 1986, has over 30 years of service in teaching and 12 years of service in University administration in various capacities. He was the Special Officer for the establishment of the Tamil

Nadu Fisheries University in 2012-13 and later served as Dean of Fisheries College and Research Institute Thoothukudi from 2013 to 2018. Academically acclaimed he won University Gold medal both at masters and doctoral programmes, respectively from TNAU, Coimbatore and UAS, Banglaore. He had his post-doctoral research at the Hroshima University, Japan and a specialized training on Molecular diagnostics from Texas A&M University, USA. He specialized in Fish Processing Technology and Fisheries Microbiology and also has keen interests in social sciences and field work.

He has successfully completed eight projects (R&D and infrastructure) as PI with an outlay of Rs 685 lakhs obtained from various funding agencies. He published more than 100 research and technical articles, 5 books, 1 book chapter, 5 teaching & 12 training manuals and 42 popular articles. Currently, he is serving as a member of the Scientific Panel on Fish and Fish Products, FSSAI, New Delhi and Chairman of the Research Advisory Committee of the ICAR-CIFT, Kochi.

Data Acquisition and Analysis in Fisheries

T. Umamaheswari
Assistant Professor (Fisheries Economics)
Department of Fisheries Extension Economics, and Statistics
TNJFU-Fisheries College and Research Institute
Thoothukudi, Tamil Nadu

S. Balasundari, M.F.Sc.
Dean
TNJFU- Dr. MGR Fisheries College and Research Institute
Thalainayeru, Tamil Nadu

G. Sugumar
Former Vice-Chancellor
Tamil Nadu Dr.J.Jayalalithaa Fisheries University
Nagapattinam, Tamil Nadu

NIPA® GENX ELECTRONIC RESOURCES & SOLUTIONS P. LTD.
New Delhi-110 034

NIPA• GENX ELECTRONIC
RESOURCES & SOLUTIONS P. LTD.

101,103, Vikas Surya Plaza, CU Block
L.S.C.Market, Pitam Pura, New Delhi-110 034
Ph : +91 11 4386 0225, 9717133558, 9540816132
E-mail: newindiapublishingagency@gmail.com
Website: www. niparesources.com

Print ISBN: 978-81-19072-29-3

ebook ISBN: 978-81-19072-17-0

Composed and Designed by NIPA®.

Preface

The sustainable management of fisheries resources is essential for ensuring the long-term viability of both marine ecosystems and the human communities that depend on them. The global demand for seafood continues to rise, placing increased pressure on fish populations and necessitating robust, accurate methods for data acquisition and analysis. This book, "Data Acquisition and Analysis in Fisheries," aims to provide a comprehensive guide to the state-of-the-art techniques and methodologies used in fisheries research today. The importance of precise and reliable data in fisheries cannot be overstated. Effective management strategies hinge on our ability to gather accurate data on fish populations, their habitats, and the environmental factors that influence them. This edited volume brings together contributions from leading experts in the field, offering a diverse range of perspectives and expertise on the various aspects of data acquisition and analysis.

From traditional methods such as net sampling and acoustic surveys to advanced techniques including satellite remote sensing, genetic analysis, and artificial intelligence, this book covers a broad spectrum of approaches. Each chapter delves into the specific methodologies, tools, and technologies employed, providing detailed explanations and case studies that highlight their application in real-world scenarios. The book also addresses the challenges and limitations associated with different data acquisition techniques, offering insights into how these can be mitigated or overcome. In addition to the technical aspects, this volume emphasizes the importance of data quality and management. Proper data stewardship ensures that collected data is accurate, consistent, and usable for long-term research and policy-making. Chapters on data validation, storage, and sharing protocols provide essential guidelines for maintaining the integrity of fisheries data.

The collaborative nature of fisheries research is a recurring theme throughout the book. Successful data acquisition and analysis often require multidisciplinary approaches and international cooperation. We are at a critical juncture in fisheries science, where the integration of traditional knowledge with cutting-edge technology can lead to significant advancements in our ability to monitor and manage fish populations sustainably. By providing a thorough grounding in the principles and practices of data acquisition and analysis, this book aims to equip researchers, managers, and students with the knowledge and tools they need to contribute to this vital field.

We would like to extend our gratitude to all the contributors for their hard work and dedication. It is our hope that this book will serve as a valuable resource for anyone involved in fisheries research and management, fostering innovation and collaboration in the pursuit of sustainable fisheries worldwide. We are deeply grateful to the numerous individuals and organizations who have made this book possible. Our sincere thanks to the contributing authors for sharing their expertise and insights, that have greatly enhanced the quality of this work. We also acknowledge the support of the faculty of the constituent units of Tamil Nadu Dr. J. Jayalalithaa Fisheries University, Nagapattinam whose commitment to fisheries research has been instrumental in the development of this volume.

Finally, we extend our appreciation to the editorial team and our publisher for their professionalism and dedication in bringing this volume to fruition. It is our collective hope that "Data Acquisition and Analysis in Fisheries" will inspire and inform the next generation of fisheries scientists and managers, contributing to the sustainable stewardship of our marine resources.

Editors

Contents

List of Contributors

1. Data Analytics in Fisheries: An Introduction
   Dr. A. K. Vasisht, ADG (PIM) (Retd), ICAR, New Delhi
2. Nuances in Project Formulation for Production, Marketing Development and Fisheries Finance,
   Dr. Shyam S Salim, Principal Scientist Fishery Resources Assessment, Economics and Extension Division (ERAEED), ICAR-CMFRI, Kochi
3. Application of Statistical Software in Fisheries
   Dr. K. B. Vedamurthy, Assistant Professor (Dairy Economics) Department of Dairy Business Management Dairy Science College, KVAFSU, Bengaluru
4. Design of Experiments in Fisheries and Aquaculture
   Dr. A. Dhandapani, Principal Scientist & Head Information and Communication Management Division, ICAR-NAARM, Hyderabad, Telangana.
5. Descriptive Statistics, Graphical Analysis, Info Graphics, Growth Rate Analysis and Data Mining Techniques in Fisheries
   Dr. Neha Wajahat Qureshi, Scientist Fisheries Economics, Extension and Statistics Division, ICAR-CIFE, Mumbai
6. Cluster Analysis for Researchable Data Sets in Fisheries
   Dr. J. Ashok Kumar, Scientist (SS), Nutrition, Genetics and Bio-technology Division, ICAR-CIBA, Chennai
7. Studying Consumer Preference of Fisheries Products: Application of Conjoint Analysis
   Dr. V. Geethalakshmi, Principal Scientist Extension Information and Statistics Division ICAR-CIFT, Cochin
8. Basics of Time Series Analysis and Forecasting Using Longitudinal Data in Fisheries
   Dr. Shinoj Parappurathu, Senior Scientist, Fishery Resources Assessment, Economics and Extension Division (FRAEED), , ICAR-CMFRI, Kochi
9. "Resource Cost Ratio" Approach to Estimate Social Profitability of Fisheries Enterprises -The Case of Integrated Fish Farming
   Dr. Biswajit Debnath, Subject Matter Specialist (SMS) ICAR-KVK, NEH region, South Tripura
10. Methodologies for Analysis Fish Marketing Data
    Dr. Biswajit Debnath, Subject Matter Specialist (SMS) ICAR-KVK, NEH Region, South Tripura
11. Analysis of Classic Breeding Experimental Data using Linear Mixed Models and Estimation of Variance Components
    Dr. J. Jayasankar, Principal Scientist & Head, Fishery Resources Assessment, Economics and Extension Division (FRAEED), ICAR-CMFRI, Kochi.

12. Linear, Non-Linear and Logistic Regression in Fisheries Data Analysis
Dr. Ramasubramanian, V., Principal Scientist & Head Research Systems Management Division, ICAR-NAARM, Hyderabad, Telangana. ram.

13. Market Integration and Price Transmission Analysis of Time Series Data in Fisheries
Dr. M. Krishnan, Principal Scientist (Retd), CIFE/ NAARM & Adviser, State Level Project Advisory Committee for Marine Resources, Andhra *Pradesh.*

14. Introduction to Big Data in Fisheries
Dr. S. Ravichandran, Principal Scientist Information and Communication Management Division, ICAR-NAARM, Hyderabad, Telangana

15. Limited Dependent Variable Models in Fisheries
Dr. V.R.Kiresur, Controller of Education, University of Agricultural Sciences, Dharwad, Karnataka.

16. Variance Components and Estimation of Genetic Parameters
Dr. Shrinivas Jahageerdar, Principal Scientist, Fish Genetics and Biotechnology Division, ICAR-CIFE Mumbai,

17. Non Parametric Data Analysis in Fisheries
Dr. P. Sethuraman Sivakumar
Principal Scientist, Section of Social Sciences, ICAR-CTCRI, Trivandrum.

18. Cutting Edge Technologies in Fisheries Data Analytics
Dr. G.R. Ramakrishna Murthy, Principal Scientist Education Systems Management Division, ICAR-NAARM, Hyderabad, Telangana.

19. Methodologies for Assessing Climate Change Impacts on Agriculture, Fisheries and Aquaculture
Dr. V. Saravanakumar, Professor (Agricultural Economics) Department of Agricultural Economics Tamil Nadu Agricultural University Coimbatore,

20. Unique Approaches in Impact Evaluation
Dr. N. Sivaramane, Principal Scientist Agribusiness Management Division, ICAR-NAARM, Hyderabad, Telangana

21. Methodological Issues in Fisheries Data Collection
Dr. R. Jayaraman, Program Director (Retd), Fisheries Business School, Vaniyanchavadi, Chennai.

22. Writing Research Papers: Breaking Myths and Building Right Perspectives
Dr. P. Krishnan, Director, Bay of Bengal Programme, Chennai

1

Data Analytics in Fisheries An Introduction

Vasisht A.K.

Introduction

The term analytics and data analytics are used inter-changeably. When referring to analytics, it means "analysis of data" or "what can be learnt from an analysis of data". In an industry like fishing industry, the researchers have been slow to adopt the tools used to portray accurate representations of data. For example, a dashboard with historical data from the past ten or fifteen years displayed on a digital map can enable the researchers to decide the places where fishing can be good. This map can help in locating the points fished during specific seasons. This will enable the researchers to filter the data on various parameters like area, year, season etc. or on any other specific parameter which can give valuable information. Similarly, this type of dashboard can affect the decision making process while fishing, like one can find the monetary value associated with each trip during a specific season or compare the values associated with two different trips to gauge the cost-benefits.

Data Analytics helps in taking most informed decisions in the fishing industry if one of the goals is to increase income while ensuring sustainability. It helps in deciding whether to reduce overhead costs or to increase expenditure to avoid overhead costs in future.

Fisheries sector undergoes continuous changes with introduction of newer technologies evolved through research and development institutions. Statistics per se deals with generation of data, data management, data analysis and information generation from data. There are varied disciplines of fisheries viz., Aquaculture, Fisheries Resource Management, Fish Genetics, Fish Biotechnology, Aquatic Health, Nutrition, Environment, Fish Physiology, and Post-harvest Technology, wherein research on practical aspects is undertaken which affect production and sustainability.

Data needs in fisheries vary according to the type of research conducted by the researcher. Biologists working on species behaviour, growth etc. require information on the spatial distribution and catch. The effect of population size on producer's cost may be studied to predict profits. In order to formulate management measures, policy makers use macro level data on infrastructure, employment, earnings, investment etc. If there are multiple and sometimes conflicting goals to be optimized like, increasing fish production, maintaining sustainability, increasing employment, increasing income of farmers etc. under various constraints and bounds, there is a need of employing Interactive Multiple Goal Linear Programming (IMGLP) by linking bio-physical parameters with socio-economics.

If there is a need to make administrative and management decisions, data needs to be collected on the production, exports, socio-economics etc. Likewise, huge amount of data is required to be collected on the infrastructure facilities, and inventories such as number of mechanized, motorized and non-motorized fishing crafts, fish landing centres, fishing harbours, types of gears and accessories, fish markets, cold storages, socio-economic data like population of fishermen, welfare schemes, co-operative societies, financial assistance, subsidies etc.

The future of fisheries depends on big data analytics. Big data offers vital tools for managing fisheries, e.g. fishing trips could be managed by-, reducing fuel consumption and crew costs. Data analytics refers to the process of examining data sets to draw conclusions about the information they contain. The techniques of data analytics enable one to extract valuable insights from raw data by uncovering various patterns. Many data analytics techniques use specialized systems and soft wares that integrate machine learning algorithms, automation and other capabilities.

Data analytics process include:- i) data collection, ii) working on data quality, iii) building the model and iv) running the model with full data. There is a difference between data analysis, data mining, and data modelling. Data analysis is done with the purpose of finding answers to specific questions. Data mining is about finding different patterns in data by way of application of various mathematical and computational algorithms to raw data. Data modelling is about how organizations organize or manage the data.

Following are the data analytics tools being used now-a-days:

1. **Python**: It is an object oriented programming language
2. **R**: This is a leading programming language for statistical modelling, visualization, and data analysis. It is mainly used by the Statisticians for statistical analysis, Big Data and Machine Learning.

3. **SAS**: This is Statistical Analysis Systems software, which is used for Business Intelligence, Data Management and Predictive Analysis.
4. **Excel**: It is a spreadsheet and powerful tool for data collection and analysis.
5. **Power Business Intelligence**: It is a powerful business analytic solution for visualization of data, connecting to many data sources and share the outcome across organization.
6. **Tableau**: It is Business Intelligence tool developed for data analysts, where one can visualize, analyse and understand data. It provides exploration of any type of data spreadsheets, data bases, data on cloud services etc.

The data analytics process has some key components. The process begins with:

a) **Descriptive Analytics:** This is the process of describing historical trends in data and involves measuring indicators like Return on Investment (ROI). However, this process does not make predictions or directly informed decisions. It focuses on summarizing data in a meaningful and descriptive way.

b) **Advanced Analytics:** It makes use of advanced tools to extract data, make predictions and discover trends. These tools include classical statistics as well as machine learning. Machine learning technologies such as neural networks, natural language processing etc. address "What if" questions. These techniques enable one to draw meaningful conclusions from complex and varied data sources.

Types of Data Analytics

There are four types of data analytics namely, descriptive, diagnostic, predictive and prescriptive analytics. Descriptive analytics requires collection of relevant data, data analysis and data visualization through which one can have insight into past performance.

Diagnostic analytics helps answer questions about why things happened. These techniques supplement basic descriptive analytics. They take findings from descriptive analytics and probe further to find the cause.

Predictive analytics answers questions about what will happen in the future. These techniques use historical data to identify trends and predict for future. Predictive analytic techniques include a variety of statistical and machine learning techniques such as neural networks, decision trees and regression.

Prescriptive analytics helps in answering questions about what should be done. Data driven decisions can be made using insights from predictive analytics. Using this analytics, one can make informed decisions in the face of uncertainty. These techniques rely on machine learning strategies in order to find patterns in large data sets. It enables estimation of likelihood of different outcomes by analyzing past decisions.

Role of Data Analytics

Data analyst's role involves interaction of information technology, statistics and business. The primary goal of a data analyst is to discover patterns in data, which in turn increases efficiency and improves performance. The steps in the data analytics process include data mining, data management, statistical analysis, and data presentation.

Data mining is one of the most time consuming step in the data analysis process. It deals with extracting data from data sources which are unstructured. The unstructured data may be in the form of written texts, large data bases or raw data. In data mining, one has to extract, transform and load data, which is then in a manageable form.

Data management, also known as data warehousing, involves designing and implementing data bases. This generally involves managing Structural Query Language (SQL) databases. SQL is used for accessing, cleaning and analyzing data that is stored in databases.

Statistical analysis is the major component of data analytics. Statistics and machine learning techniques are used to analyse data. There are models in statistics that reveal trend in data and these can be used for making predictions and make sound decisions on the basis of that.

Data presentation is the last step in most data analytic process. This helps in sharing information with the stakeholders.

Ways to use Data Analytics

The researcher can achieve lot of values for his/her research from data. But for this, analytic component is needed. The main benefits of data analytics are:

a) **Improved decision making :** Informed decisions lead to better outcomes. A lot of guess work from planning gets removed by data analytics. Modern technology helps in collecting data on a regular basis and keep on analyzing it in order to update ones knowledge and get insights as changes take place.

b) **More effective marketing :** For effective marketing, fine tuning of targets is necessary, which help in getting optimal outcomes. Data analytics helps in getting useful insights on various aspects.

c) **More efficient operations :** In order to have efficient processes and save resources, one can employ the tools of data analytics by collecting different forms of data from different sources and thus one can improve targets and make better predictions. For example, information on consumer behaviour can help in furthering market research and for this purpose data needs to be compiled.

Data Analytics Technology

It is possible to get much deeper data insights using ever growing volume of data and advanced analytics technology. These insights are more accurate and extensive in nature. The current data can be used for taking immediate decisions and also help in taking proper future decisions.

The important technologies which a data analyst use are:

a) Machine Learning

b) Data Management

c) Data Mining

d) Predictive Analytics

Machine learning is a sub set of Artificial Intelligence (AI). AI is the field of simulating human intelligence using computer technology. Machine learning helps the applications to take in data and analyse it to predict outcomes without programming the system.

One of the pre-requisites for analysing data is to possess organized data and for this there has to be proper system to manage continuous flow of high quality data and stored in a data repository for usage as and when needed.

Data mining is the process of sorting huge amounts of data to identify patterns and figuring out relevant relationships. This can be used to analyse and take informed decisions. The data mining technology can quicken the process of accomplishing desired tasks.

Predictive analytics helps in predicting future outcomes based on historical data and also gauge the possibilities of occurrence of different outcomes. Accurate predictions result in better decisions. The algorithms used by these technologies are statistical and machine learning.

Advantages of Data Analytics

1. Data analytics has the ability to make faster, more informed decisions backed by facts.
2. It helps in getting deeper understanding of requirements in varied disciplines.
3. By using data analytics, one can implement preventive measures by having increased awareness of risks.

The increased availability of data through sensors, satellite imagery, cameras, drones and other technologies make advanced data analytics more viable. There are better tools for communicating information such as smartphones and Internet of Things (IoT). The technological advances in the areas of artificial intelligence, data storage, computational power etc. result in enhanced capabilities of handling data and analyse information using sophisticated algorithms. There are image recognition and object-detection tools which can provide fishermen with important information on the content of catch in real time, including species, volume, and fish size. It has been very easy now-a-days to transmit data for analysis from vessels by making use of IoT, land and satellite based module networks, and smartphones.

There are many technology-enabled processes like fisheries management, detection and capture, catch sorting, reporting and surveillance & control which can be made use of in analytics. In fisheries management, data analytics helps in shifting to a data-rich environment that provide reliable assessments. It also helps in continually assessing fishing stocks. The detection of fish over the entire fishing territory can be more reliable by using high resolution models and daily forecasts using advanced forecasting techniques. The advanced analytics enable automatic scanning and control of sea food product quality through cameras and intelligent sorting systems.

There have been reporting mechanisms by recording captured species via log books. But with advanced analytic tools, reporting can be done using artificial intelligence recognition software.

The surveillance and control on fishing activities can be more reliable by using decentralized information management system involving little human intervention.

Data analytics has also a greater role in improving the detection of fish. The issues relating to fish aggregation and migration, which change in response to temperature, wave height, presence of sea ice, and other ocean conditions, can be monitored through satellite imagery. The advances in this field can help determine the distribution and migratory patterns of target species over time & space with greater accuracy and frequency.

There are web-based tools like "Detect IT: Fish" and "Google's Global Fishing Watch", which use Big Data Analytics to identify illegal fishing activities in a given area. Thus, it helps in increasing transparency in global fish trade, curb illegal, harmful and dangerous fishing practices.

References

Advanced analytics: What it is, Techniques and Tools,www.looker.com
Big data analytics to fight challenges of aquaculture and fisheries sector, www.researchgate.net
https://www.directionsmag.com/pressrelease/7632
https://www.northeastern.edu/graduate/blog/what-does-a-data-analyst-do/
The next step for the fishing industry: Data Analytics, www.fishermensnews.com, 2018.
What is advanced analytics, www.sisense.com
What is data analytics? – 4 ways to use data analytics, www.lotame.com

2

Nuances in Project Formulation for Production, Marketing Development and Fisheries Finance

Shyam S. Salim

Introduction

Indian fisheries and aquaculture is an important sector of food production providing nutritional security, besides livelihood support and gainful employment to more than 14 million people, and contributing to agricultural exports. With diverse resources ranging from deep seas to lakes in the mountains and more than 10% of the global biodiversity in terms of fish and shellfish species, the country has shown continuous and sustained increments in fish production since independence. The sector constitutes about 6.3 percent of the global fish production and contributes to 0.91 percent of the GDP and 5.23 percent of the agricultural GDP. The total fish production of 13.34 million metric tonnes presently has nearly 74 percent contribution from the inland sector and rest by the marine sector. Paradigm shifts in terms of increasing contributions from inland sector and further from aquaculture have been significant over the years. With high growth rates, the different facets, viz., marine fisheries, coastal aquaculture, inland fisheries, freshwater aquaculture, and coldwater fisheries are contributing to the food basket, health, economy, exports, employment and tourism of the country. More than 50 different types of fish and shellfish products are being exported to 75 countries around the world. Fish and fishery products have presently emerged as the largest group in agricultural exports from India, with 13.77 lakh tonnes in terms of quantity and ₹ 45,106.89 crore in value. This accounts for around 10% of the total exports and nearly 20% of the agricultural exports, and contribute to about 0.91% of the GDP and 5.23% to the Agricultural GDP of the country.

The Pradhan Mantri Matsya Sampada Yojana (PMMSY) is an initiative launched by the Government of India to establish a comprehensive framework and reduce infrastructural gaps in the fisheries sector. The scheme was announced by the Finance Minister, Smt. Nirmala Sitharaman during her speech in the Parliament of India while presenting the Union budget for 2019–20 on 5 July 2019. The government intends to place India in the first place in Fish production and processing by implementing Blue Revolution. This scheme is in line with government's aim to double the farmers' income by 2022–23.The policy envisages integrating all the fishermen with agricultural farmers and providing all the facilities available through various farmer welfare schemes to the fishermen. A new dedicated Department of Fisheries was constituted in a newly carved out Ministry of Fisheries, Animal Husbandry and Dairying to implement this and other policy initiatives of the government.

Projects - The Cutting Edge of Development

Post liberalization era of Indian economic scenario reveals disinvestments of public enterprises and the flourishing of private firms. Bank rates were slashed in order to boost the morale of venture capitalists. Fisheries with an annual sectorial growth of 8 percent became an ideal niche for bulk of those investors who lost their interest due to BCR from agriculture. Family enterprises flourished in many places all over India and entrepreneurship became a key word in fisheries sector.

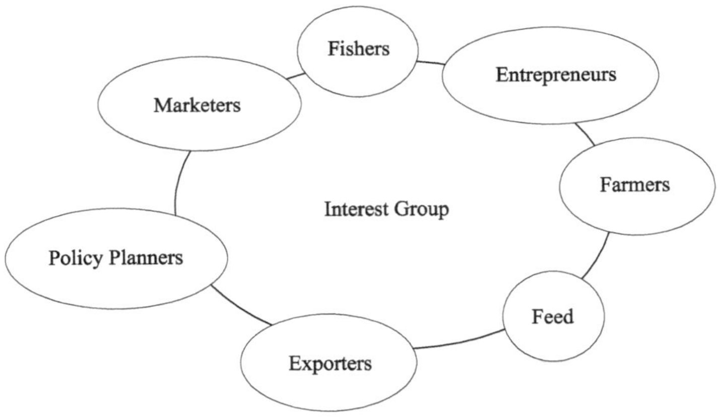

Fig. 1. Interest groups in fisheries projects

Project is an investment activity in which financial resources are expended to create capital assets that produce benefits over an extended period of time. That's why projects are often referred to as the cutting edge of development. Project preparation is clearly not the only aspect of fisheries development or planning. Identification of national fisheries development objectives, selecting priority areas for investment, designing effective price policies, and mobilizing resources are all critical. Unless, projects are carefully prepared in substantial detail, inefficient or even wasteful expenditure is almost sure to result a tragic loss in nations short of capital.

Project formulation is an important part of entrepreneurship development. Fisheries sector, hosting a gamut of entrepreneurship venture, corroborates the importance of technical know-how in project planning, formulation and implementation. Diverse information in projects targeted at a heterogeneous group has been modified to suit the needs of an entrepreneur seeking assistance in his venture.

Project Funding in Fisheries

Foreseeing the vast resource potential and possibilities in the fisheries sector, a separate Department of Fisheries was created in February, -2019. The Government has merged all the schemes of fisheries sector into an umbrella scheme of 'Blue Revolution: Integrated Development and Management of Fisheries' focusing on increasing fish production and productivity from aquaculture and fisheries resources, both inland and marine. Towards realization of these objectives, the creation of the Fisheries and Aquaculture Infrastructure Development Fund (FIDF) was approved with a total fund size of Rs 7,522.48 crores.The project funding opportunities has become diverse over the years and is given in Fig. 2.

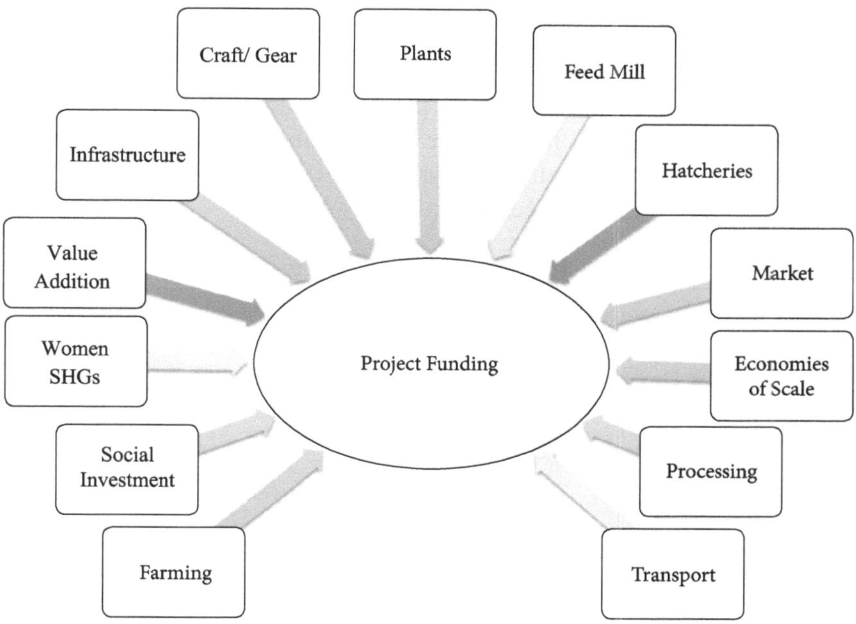

Fig. 2. Diversification in project funding

Fisheries Project Environment

The fisheries project environment is intricate with different domains (Fig. 3) operating in unison and each requiring specific choices in decision making.

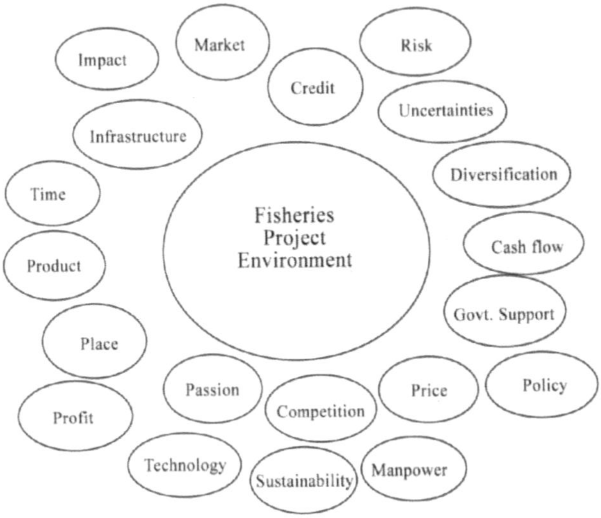

Fig. 3. Fisheries project environment

Project Cycle

There is a natural sequence by which projects are planned and carried out and this sequence is called 'Project cycle'. International development agencies tend to use the World Bank methodology [Baum, 1982]. Under this scheme the cycle is broken down to six stages given in Fig. 4.

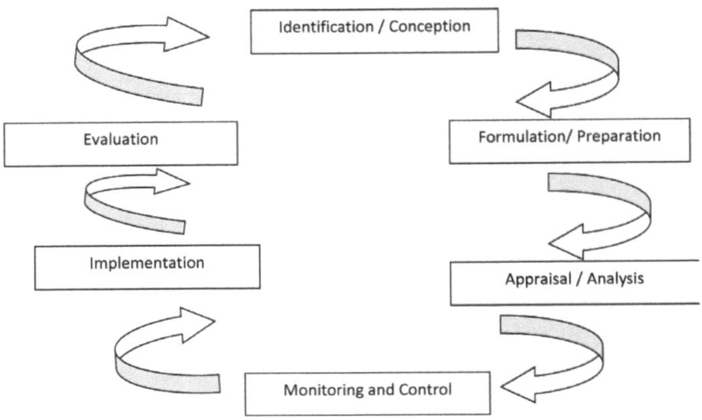

Fig. 4. Project cycle

Investment feasibility assessment of fisheries projects

In order to ascertain whether an aquaculture investment project is feasible or not, a cooperative evaluation should first be conducted by both the biologist and economist. Only those species and projects that are suited to the local environment and biologically feasible for development should be considered. Thereafter, a socio-economic study can be undertaken. Basically, an economic evaluation includes both the production and marketing functions.

a) The first requirement for any aquaculture investment project in both the public and private sectors is the availability of suitable land and water resources.

b) The species selected for development should be adapted to the local environmental conditions and the stocking materials and suitable feed should be readily available at reasonable cost. The species should also be fast growing and culture technology should be locally available.

c) There should be no legal constraints on development (this is particularly important for private investors).

d) The products of the investment project should have a high market demand with a reasonable price.

e) The investment project should be financially lucrative compared to other investment opportunities for private investors and should also be socio-economically feasible with alternative means of achieving the national objectives for public investment. Private investors usually use profitability as a measure of financial feasibility when assessing commercial aquaculture projects, and public officials usually consider socioeconomic benefit-cost and/or the social internal rate of return as measures of economic feasibility along with some qualitative judgments.

   In order to evaluate the feasibility of an investment protect in aquaculture, one must consider six criteria:

   i. Resource availability,

   ii. Environmental suitability,

   iii. Biological feasibility,

   iv. Market potential,

   v. Economic feasibility, and

   vi. Institutional feasibility.

f) It is also important to realize that many variables ought to be considered for each criterion. Each variable can be assessed as favourable, partially favourable, unfavourable, etc. Each ranking can then be scored (or coded) numerically weighted or un-weighted. Next, a general score or code can be assigned to each criterion after evaluation of all the sub scores and codes, and the bio-economic feasibility can be determined by weighting the general score or code for each criterion. This procedure can be varied to suit particular projects.

Criteria for selection of fishery projects

The important criteria for the selection of fisheries project are briefly summarized below:

i. **Work selection criteria:** This relates to immediate needs of the project area and has a direct or indirect relation in increasing prospects of fisheries production, income and employment.

ii. **Priority criteria:** It deals with whether the project implemented falls under priority area or not.

iii. **Social criterion:** It considers the direct employment prospects, ecological balances, externalities, pollution etc.

iv. **Financial criteria:** According to this criterion, it is determined whether the required amount of capital is supplied or not for the implementation of the project. In case the execution is delayed, additional capital requirements are to be assessed.

v. **Supply criterion:** This is concerned with available resources like physical inputs, labour availability and other resource endowment. Supply of skilled labourers and un-skilled labourers and technical personnel are to be evaluated for the completion of the project on time.

vi. **Implementation criterion:** This is based on organizational and management abilities of technical personnel. Organization refers to the organization hierarchy of the implementing agency. The availability of staff at various cadres, demarcation of authority and linking of authority and responsibility etc. are to be analysed in detail.

vii. **Project benefits criterion:** Both tangible and intangible benefits must be correctly assessed and evaluated. In this process, the benefits accrued due to forward linkages and backward linkages need to be given specific weightage. The benefit must address community defined priority needs in the areas of health, education, training and income generation.

Financial performance

The profitability of a fisheries project influences its value and the amount of income it generates for its producers/ beneficiaries. The financial performance of an enterprise is analysed by working out various types of indicators as given below (Fig. 5). These measures compute either in terms of undiscounted or discounted – mainly taking into consideration of the time value of money.

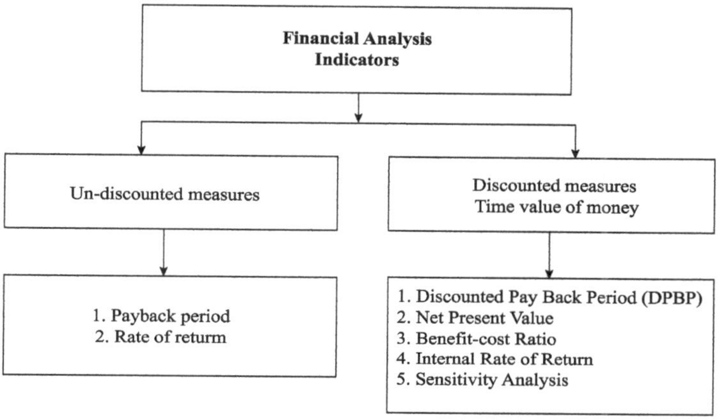

Fig. 5. Financial performance indicators

Time value of money

Many economic decisions including fish production involve benefits and costs that are expected to occur at future time period. The construction of ponds race ways, and fish tank, for example, requires immediate cash outlay, which with the production and sale of fish, will result in future cash inflows or returns .In order to determine whether the future cash inflows justify present initial investment, we must compare money spent today with the money received in the future.

The time value of money (TVM) is the concept that money you have now is worth more than the identical sum in the future due to its potential earning capacity. This core principle of finance holds that provided money can earn interest, any amount of money is worth more the sooner it is received. The time value of money (TVM) is an important concept to investors because a dollar on hand today is worth more than a dollar promised in the future. Provided money can earn interest, this core principle of finance holds that any amount of money is worth more the sooner it is received. The time value of money is determined by the following:

1. Number of time periods involved (months, years)
2. Annual interest rate (or discount rate, depending on the calculation)
3. Present value (what you currently have in your pocket)
4. Payments (if any exists; if not, payments equal zero)

5. Future value (the amount you will receive in the future. A standard mortgage will have a zero future value, because it is paid off at the end of the term).

The intrinsic value of money diminishes over time and for the same output we might need to pay more money. The intrinsic loss is captured by **Discounting** factor which is the process of finding the present value of a future payment is called discounting. The future value must be discounted to reflect the earnings lost by not being able to immediately invest the future sum in the alternative investment. The discounting factor could be rate of interest/ inflation level.

The financial feasibility analysis is done using the following capital budgeting techniques with appropriate assumptions on the duration of the farming, annual days of operation, inflation of costs and returns and related parameters. Three indicators will be estimated namely, Net Present Value (NPV), Benefit Cost Ratio (BCR) and the Internal Rate of Return (IRR).

There are 4 discounted measures:

i) Discounted Pay Back Period (DPBP)

Number of years required to return the capital investment, which is computed by the cumulative sum of discounted cash inflows.

ii) Net Present Value / Net Present Worth (NPV)

NPV is the difference between the discounted benefit and discounted cost where present value of the expected future net cash flows is discounted at a specified discounted rate.

$$\text{NPV} = \sum_{t=1}^{n} \frac{CF_t}{(1+r)^t}$$

$$= \left[\frac{CF_1}{(1+r)^1} + \frac{CF_2}{(1+r)^2} \ldots\ldots\ldots\ldots\ldots\ldots\ldots + \frac{CF_t}{(1+r)^t} \right] - C,$$

where, C is the Capital investment, (CF_1 ....CF_t) is the Cash inflow and t is the time in years

iii) Benefit Cost Ratio (BCR)

Ratio of the discounted benefit of cash inflows to the discounted investment outlay.

$$\text{B / C ratio} = \frac{\text{Total Discounted Benefit}}{\text{Total Discounted Cost}}$$

$$\text{BCR} = \frac{\sum_{t=1}^{n} \frac{B_t}{(1+r)}}{\sum_{t=1}^{n} \frac{C_t}{(1+r)^t}}$$

where B_t is the Cash inflow and C_t is the Cash outflow

iv) Internal Rate of Return (IRR)

IRR is the discount rate at which Net Present Value (NPV) is equal to zero and benefit cost ratio is equal to one where discount rate is that equates the present value of the expected future cash flows or receipts to the initial cost of the project.

IRR= NPV = 0

For working out IRR two discount rates, one of which gives positive NPV and the other gives negative NPV are generally used. The formula used is as follows:

$$\text{IRR} = r_L + [\frac{P_v - C}{\Delta P_v}] \times \Delta r$$

where, IRR is the Internal Rate of Return, r_L is the lower rate of discount, P_V is the present value at lower value of discounting, C is the capital investment, ΔP_V is the Difference in the present values at the two discount rate, Δr is the Change in discount rate.

Risk and Uncertainties

In project analysis, once the stream of costs and benefits for a project is defined in the form of cash flows, the next step is to assess the worthwhileness of the project using a variety of appraisal criteria (or evaluation methods). *A project -manager will search for the project that will satisfy the project goals, and that includes the wisest and best use of resources to satisfy human wants. The most common criteria employed in determining the financial desirability of investment projects are NPV, BCR and IRR.* One of the real advantages of careful economic and financial analysis in fisheries project is that it may be used to test what happens to the earning capacity of the project if events differ from guesses made about then in planning for e.g. a disease outbreak, fall in prices, natural calamities like floods, etc. These unforeseen incidents can be grouped into two i.e. risks and uncertainties.

1. Risk

Few management decisions are made under conditions where the outcomes associated with each possible course of action are known with certainty. Most major managerial decisions are made under conditions of uncertainty. The frequency of uncertainty in managerial decisions and the risk involved dictate risk analysis be given due consideration in farm and project management decisions. Risk refers to the possibility that some unfavorable event will occur. The probability of a risk occurring can be quantified. It is the possibility of loss, injury, or exposure to harm. In aquaculture, risk comes from stock losses. Anything, which disrupts the rearing of fish, is likely to jeopardize production and marketing of the final product. The levels of risk vary among species and at different stages of production. The relative lack of knowledge of fish biology in comparison to some land animals makes fish production more risky than the production of food animals. As Secretan (1988) indicates, on a scale of 1 to 100, we know 75 percent of the biology of human beings, and perhaps 50 to 60 percent of the biology of chickens, cows, pigs and other farm animals, but only about 20 percent of the biology of aquatic species. There are numerous risks involved in the breeding, hatching and growing of aquatic organisms under intensive management systems.

What sort of risks plagues the fisheries/ aquaculture industry? Risks may be classified into main groups: (1) socio-economic or business risk and (2) physical or pure risks.

Socio-Economic / Business: Social aspects of socio-economic risks include changes in tastes, attitudes, or social behavior towards production and consumption of a certain species. The expansion of aquaculture depends on individuals changing their attitudes towards species cultured under intensive closed systems. This may be done through government programs, advertising, and public relations. For example, changes in consumer purchases of catfish have been achieved through advertising and public relations. The growing popularity of catfish may be stifled, however, if "off flavour" problems continue to plague the industry.

Economic risks: Economic risks such as changes in price of inputs and output, inflation, recession, depression and other economic conditions that affect national income are primary concerns of commercial fish producers. As demand lags behind supply, producers are concerned that prices will fall. This is presently the case in the U.S. catfish industry. Producers are being warned that they should secure markets before expanding production. Also, the degree of elasticity with respect to supply and demand at both the farm and processor level is a clue to the level of economic risks associated with fish production. Processors facing a more inelastic demand than producers will tend to be less concerned about demand lags. This is one reason that producers are beginning to favour more producer associations or cooperative type marketing organizations.

Marketing risks: Risks may also result from uncertainty in demand, supply and prices. When to move the product to market is the age-old nemesis of farmers. Fish farmers are no different. Significant seasonal price level differences exist in many aquaculture product markets. Today, more farmers in colder climate are over wintering fish to try to market them when there is less supply available to consumers. Additionally, new technologies and product forms are being evaluated in an attempt to avoid some of the marketing risks. Smoked fish as well as dried, frozen, or canned fish are forms used in various markets to reduce the risks associated with marketing time. Assume that forecasters are overly optimistic in their estimates of prices and consumer demand. This optimism is likely to encourage farmers to intensify production (higher stocking rates) in the short-run and expand production (more ponds) in the long run. Intensification increases the potential for diseases, problems such as "off-flavor", and other environmental concerns. The fish arrive at the market only to remain unsold because of weak consumer demand resulting from a dislike for the quality of the fish on the shelf, or insufficient income to purchase fish and other market foods. Longer-term expansion of production means greater amounts of capital and land committed to the aquaculture practice. Because ponds are much easier to build than to remove, these commitments tend to become irreversible, even if prices decline. Once again market conditions dictate difficulties for the producer.

Production risks: Many of the marketing risks are also related to production problems. Marketing problems may be logistical in nature, which may impede production schedules. The timely supply of fingerlings may affect the quantity of food fish produced at a given time. This may result in grave financial problems for producers. Production risks may also be due to lack of trained manpower to manage the operation. This results in serious constraint or even failure in any aquaculture enterprise.

Other risks: Other socio-economic risks encountered are financial and political. Financial risks relate to changes in supply of funds for production and marketing. Credit restriction and availability often affect the aquaculture industry. Lack of education and understanding of aquaculture production processes among lenders is common in areas where the industry is developing. Political risks affect not only an enterprise, but the whole sector. Changes in government and governmental policies have been known to cause changes in supply and demand of inputs and fish. Governmental regulations may affect all stages and aspects of the industry. Regulations on feed, import of inputs, the introduction of species, and changes in labour laws may greatly influence the industry.

Physical or pure risks: Physical risks results from conditions of nature, such as rain windstorms, clouds, flooding, and drought. Other types of pure risks are plant breakdowns, and failure of safety and other devices. These risks associated with physical or pure risks can be managed to minimize their effects on producers.

2. Uncertainty

It is a situation in which the probability of an outcome is not known. Insurance cannot provide any cover against uncertainty. Uncertainty is a state of being doubtful about future events, which cannot be foreseen exactly.

Types of Uncertainty

Price uncertainty: It is associated with the price of products and input factors, such as price of fish in a market.

Yield uncertainty: The fluctuations in yield are associated with weather conditions and incidence of diseases and pests and the impact of new practices.

Technological uncertainty: Technological changes influence production function and create conditions of variability, which, in turn, lead to uncertainty.

Institutional uncertainty: Conditions of tenure, functioning of credit agencies, action and outlook of farmers are examples of institutional uncertainty.

Risk and Uncertainty Averting Mechanisms

- **Diversification:** Diversification is a technique that reduces risk by allocating investments among various financial instruments, industries, and other categories. It aims to maximize returns by investing in different areas that would each react differently to the same event. Crop diversification provides the fish farmers with a wider choice in the production of a variety of cultivable species in a given area so as to expand production related activities on various fish species and also to bring down the possible risk.
- **Crop insurance:** The crop insurance policies provide coverage for loss of production/yield or how much a farm produces. Some plans combine yield and price coverage. They cover loss in value due to a change in market price during the insurance period, in addition to the perils covered by the standard loss of yield coverage.
- **Continuous or Sequential Marketing:** Aimed at continuous marketing of the product through the different advertisement so that the market penetration is improved.

- **Future Production Contracts:** A future production contract is a legal agreement to buy or sell a particular commodity asset, or security at a predetermined price at a specified time in the future. Future contracts are standardized for quality and quantity to facilitate trading on a futures exchange.
- **Government Programs:** The government programmes include stabilisation policies which insulates the different risk and uncertainties in terms of loan waiver, evasion of taxes, providing subsidies.
- **Third-Party Equity Capital** - Capital received for an interest in the ownership of a business which reduces to the risk of the owner for the investment or liability which he/ she possess.

3. Sensitivity Analysis

Sensitivity analysis employs a simple technique to assess the effects of adverse changes on a project. It involves changing the value of one or more selected variables and calculating the resulting change in the NPV or IRR. The extent of change in the selected variable to test can be derived from post evaluation and other studies of similar projects. Changes in variables can be assessed one at a time to identify the key variables. Possible combinations can also be assessed. Sensitivity analysis should be applied to project items that are numerically large or for which there is considerable uncertainty. The results can be presented together with recommendations on what actions to take or which variables to monitor during implementation and operation. The merits includes forces management to identify the underlying variables and their relationships, shows how robust or vulnerable a project is to change in underlying variables and indicates the need for further work in terms of gathering information in NPV or IRR is highly sensitive to changes in some variables. The demerits includes inability to provide leads - if sensitivity analysis merely presents complicated set of switching values, it may not shed light on the characteristics of the project and the fact that the impact of variation in one factor at a time, holds other factors constant, may not be very meaningful when underlying factors are likely to be inter-related.

Methodology

Sensitivity analysis can be done to ascertain the project feasibility at three different stages.

i) **Increasing cost of capital or interest rate increases:** The increasing cost of capital or the interest rate increases can be accounted in the sensitivity analysis by computing the NPV and BCR at different discount rates and thereafter checking the profitability of the changes.

ii) **Escalation of cost of the project due to different risks involved:** The cost of the projects gets escalated due to the various risk factors involved in the business which include the prophylactic measures needed to control and prevent the disease outcome, application of more fertilizers than the expected, more number of irrigations, more number of man days increase due to the inefficiency of human labour, etc. This increase the cost of the project can be accounted by the ex ante approach of increasing the project cost by 10 percent and 20 percent and later working the NPV and BCR with the benefit stream keeping unchanged.

iii) **Uncertainties resulting due to differences in the price receivables :** The uncertainties in the project benefit stream arise due to the uncertain nature of the prices that are expected in the market after the harvests. The uncertainties are basically due to the reason that the factors determining prices itself are subjected to changes. The other uncertainties include the yield uncertainty, technological uncertainty and institutional uncertainty. In countering the uncertainties, the anticipated benefit stream in the project can be reduced by 10, 20, 30 percentages and the NPV and BCR are computed accordingly, keeping the project cost unchanged.

Example: Financial Performance

Compute the discounted payback period, net present worth, benefit-cost ratio and internal rate of return for the aquaculture project. The discount rate is 12 per cent.

Year	Cash outflow	Cash inflow	Discounting factor	Discounted cash outflow	Discounted cash inflow
1.	250000	-	1	250000	-
2.	-	120000	0.8929	-	107150
3.	-	100000	0.7972	-	79720
4	-	80000	0.7118	-	56940
5	-	54000	0.6355	-	34320
Total	**250000**	**354000**		**250000**	**278130**

Solution

1. Discounted Pay Back Period = 4.5 years
2. Net Present Value = 27,8130 - 25,0000 = 2,8130
3. Benefit Cost Ratio = 27,8130 ÷ 25,0000 = 1.112

4. Internal Rate of Return

Year	Cash outflow	Cash inflow	Discount-ing factor at 12%	Discount-ed cash outflow	Discounted cash inflow at 12%	Discount-ing factor at 20%	Discount-ed cash outflow
1.	250000	-	1	250000	-	1	-
2.	-	120000	0.8929	-	107150	0.83300	100000
3.	-	100000	0.7972	-	79720	0.6944	69440
4.	-	80000	0.7118	-	56940	0.5787	46300
		250000			278130		241780
Net Present Value =					**+ 28130**		**- 8220**

There are discount rates which gives a positive and negative net present worth.

$$IRR = rL + \frac{[P_V - C]}{\Delta P_V} \times \Delta r$$

$$IRR = 12 + \frac{[278130 + 250000]}{(278130 - 241780)} \times (20 - 12)$$

$$IRR = 12 + \frac{[2813]}{(3635)} \times 8$$

IRR = 18.19

Conclusion

The project is feasible because the NPV is greater than zero, BCR is greater than one and IRR is greater than the opportunity cost of capital.

Example on Sensitivity Analysis

For the following fisheries project, perform the sensitivity analysis for the three different cases.

i) Increasing cost of capital.

ii) Increased cost of project due to risks involved at 10 and 20 percent cost like.

iii) Uncertainties due to the differences in the price receivables at 10, 20 and 30 percent reduction for the yield.

Example: Sensitivity Analysis

CASE I: Increasing Cost of Capital

Year	Cost	Benefit	d.f. at 12%	d.c. at 12%	d.b. at 12%	d.f. at 20%	d.c. at 20%	d.b.at 20%	d.f.at 25%	d.c. at 25%	db. at 25%
1	25000	0	1	25000	0	1	25000	0	1	25000	0
2	5000	20000	0.893	4465	17860	0.833	4165	16660	0.8	4000	16000
3	5000	20000	0.797	3985	15940	0.694	3470	13880	0.64	3200	12800
4	5000	20000	0.712	3560	14240	0.579	2895	11580	0.512	2560	10240
5	5000	20000	0.636	3180	12720	0.482	2410	9640	0.41	2050	8200
6	5000	25000	0.567	2835	14175	0.402	2010	10050	0.328	1640	8200
				43025	74935		39950	61810		38450	55440
				NPV	31910		NPV	21860		NPV	16990
				BCR	1714		BCR	1.5471		BCR	1.4418

Remarks

The computation of the NPV and BCR at different cost of capital indicates that the project is feasible and profitable even at 25 percent discount rate. At 25 percentage discount rate also there exists a positive NPV and BCR of more than one. The exercise indicates the high yielding capacity of the project even at higher discount rates.

CASE II: Escalation of the cost of the project due to the different risks involved

Year	Cost	Benefit	d.b. at 12%	d.c. at 12%	d.b. at 12%	Cost increase by 10%	d.c. at 12%	d.b. at 12%	Cost increase by 20%	d.c. at 12%	db. at 12%
1	2500000	0	1	25000	0	27500	27500	0	30000	30000	27500
2	5000000	20000	0.893	4465	17860	5500	4911.5	17860	6000	5358	4911.5
3	500000	20000	0.797	3985	15940	5500	4383.5	15940	6000	4782	4383.5
4	500000	20000	0.712	3560	14240	5500	3916	14240	6000	4272	3916
5	500000	20000	0.636	3180	12720	5500	3498	12720	6000	3816	3498
6	500000	25000	0.567	2835	14175	5500	3118.5	14175	6000	3402	3118.5
				43025	74935		47327.5	74935		51630	47327.5
				NPV	31910		NPV	27607.5		NPV	-4302.5
				BCR	1.741		BCR	1.58		BCR	0.92

Remarks

On increasing the cost of the project taking into consideration the different risks involved, the computed NPV and the BCR values indicate that the project is feasible and economical to a discount level rate of less than 20 percentage cost increase. At 20 percentage increase in the total cost of the project, the NPV appears to be negative and the BCR is lesser than one that are negative indicators of project appraisal.

CASE III: Uncertainties resulting due to the differences in the price receivables

Year	Cost	Benefit	DF at 12%	DC at 12%	DB at 12%	Reduction in benefit of 10%	Discount-ed benefit	Reduction in benefit of 20%	Discount-ed benefit	Reduction in benefit of 30%	Discounted benefit
1	250000	0	1	25000	0	0	0	0	0	0	0
2	5000	20000	0.893	4465	17860	18000	16074	16000	14288	14000	12502
3	5000	20000	0.797	3985	15940	18000	14346	16000	12752	14000	11158
4	5000	20000	0.712	3560	14240	18000	12816	16000	19392	14000	9968
5	5000	20000	0.636	3180	12720	18000	11448	16000	10176	14000	8904
6	5000	25000	0.567	2835	14175	22500	12757.5	20000	11340	17500	9922.5
				43025	74935		67441.5	84000	59948	73500	52454.5
				NPV	31910	NPV	24416.5	NPV	16923	NPV	9429.5
				BCR	1.71	BCR	1.56	BCR	1.39	BCR	1.21

Remarks

The uncertainties in the project benefit stream can be sensitised by the *ex ante* approach of reducing the anticipated project benefit stream at 10, 20, 30 percentages. The computed NPV and BCR ratios indicate that the project can withstand uncertainties to the tune of even 30 percent reduction in the yield due to the different uncertainties. The NPV and BCR at 30 percentage reduction in the yield in the project benefit stream were found to be ₹ 9,429 and 1.21, respectively.

Conclusion

The above tools namely cost and return indicators, budgeting analysis and financial feasibility indicators including sensitivity analysis provides various opportunities in assessing the economic efficiencies – *ex ante* and *ex post.* However, it is important to include social and environmental assessment too so as to ensure that the business option is inclusive and sustainable. The sustainable farming would encompass the facets of ecologically sound, environment friendly, socially acceptable, technologically sound, financially feasible, economically viable, user friendly, culturally compatible, indigenously resource sourced and market driven.

References

Battese G.E. and T.J. Coelli, 1995. A model for technical inefficiency effects in astochastic frontier production function for panel data. Empirical Economics, 20:325-332.

Battese, G.E. and T.J. Coelli, 1992. Frontier production functions, technical efficiency and panel data with application to paddy farmers in India, 1992. The Journal of Productivity Analysis, 3:153-169.

Coelli, T., 1996. A guide to FRONTIER version 4.1: A computer program for stochastic frontier production and cost function estimation, CEPA Working Paper 96/08, University of New England.

Gopakumar, G., K.R. Manmadhan Nair and V. Kripa., 2007. Mariculture research in India status, constraints and prospects. In: Status and Perspectives in Marine Fisheries Research in India. CMFRI, Cochin. Pp. 316-361. ISBN 81-901219 6-0.

Narayanakumar, R and Krishnan, M., 2013. Socio-economic assessment of seaweed farmers in Tamil Nadu - A case study in Ramanathapuram District. Indian Journal of Fisheries, 60 (4). pp. 51-57.

Narayanakumar, R and Shyam, S Salim and Shinoj, P and Ramachandran, C., 2017. Methodological tools for socio-economic and policy analysis in marine fisheries. CMFRI Special Publication (127). Central Marine Fisheries Research Institute, Kochi.

Narayanakumar, R., 2009. Economic analysis of cage culture of Sea bass. In: Course manual: National training on cage culture of Sea bass. Imelda, Joseph and Joseph, VEdwin and Susmitha, V, (eds.) CMFRI & amp; NFDB, Kochi, pp. 120-122.

Pandey, S. K., Shyam, S Salim and Biradar, R. S., 2005. Economics of Freshwate Aquaculture Technologies. Central Institute of Fisheries Education, Mumbai.

Rao, G Syda and Narayanakumar, R and Imelda, Joseph., 2009. Successful Open Sea Cage culture demonstration at Balasore, Orissa: A New Horizon in Mariculture CMFRI Newsletter (123). pp. 3-6.

SAS Institute Inc. 2014. SAS/ETS® 13.2 User's Guide. Cary, NC: SAS Institute Inc.

Sathiadhas, R. and R. Narayanakumar, 2010. Economic analysis of Open sea cage farming. In Course Manual. Training Programme on Sea cage farming of sea bass and lobsters, CMFRI Kochi.

Shanmugam, K.R., and Atheendar Venkataramani, 2006. Technical efficiency in agricultural production and its determinants: An exploratory study at the district level. Ind. Jn. of Agri. Econ. Vol. 61, No. 2, April-June 2006.

Shyam, S Salim and Biradar, R. S. and Pandey, S. K., 2004. Economic analysis of fisheries projects. Central Institute of Fisheries Education, Mumbai.

Shyam, S Salim and Biradar, R. S. and Pandey, S. K., 2005. Fisheries Economics and Marketing-An introduction. Central Institute of Fisheries Education, Mumbai.

Shyam, S Salim, 2002. Business Strategies for venturing into Fisheries enterprises [Teaching Resource].

3

Application of Statistical Software in Fisheries

Vedamurthy K.B.

Need of Analytics in Fisheries Management

New technologies are introduced for industrial vessels every day: Fuel and gear sensors, cameras, electronic catch reports etc. There is a global push to track small-scale fishing vessel. In the coming years, fisheries monitoring centers will have to adjust from tracking a few hundred vessels to tens of thousands of vessels, meaning huge quantity of data. To make sense of all this information, it must be properly analysed. The future of fisheries depends on big-data analytics

This may help in many ways. Fishing trips could be tailored to meet quotas in the shortest time reducing fuel consumption and crew costs. Real-time catch data could help administrations close a fishing zone once quota is reached. Fisheries administrations can monitor catch efforts over an entire region and decide which zones to close, improving management of stocks and fishing licenses. Fighting the challenge of Illegal and Unreported fishing contributing to over-exploitation of fish stocks. Use of disruptive technologies in fisheries and aquaculture like block chains, sensors and automatic identification systems (AIS) has the potential to change the processes profitability and sustainability of the sector

Globally used popular Data Science Software

- MS-EXCEL
- SPSS (IBM)
- STATA
- MINITAB
- SAS
- R
- PYTHON

MS-Excel, SPSS, STATA, SAS are proprietary softwares while, R and Python are open source softwares. Open sources have their advantages. The differences between these softwares are given below.

Open Source Software	**Proprietary Software**
Software developed and tested by open collaboration	Software is solely owned by the individual or organisation that developed it
Anyone with academic knowledge can access, inspect, modify and redistribute the software	Only the owner or publisher has the legal right to source code and access it
It gives better flexibility and more freedom for innovation	Limited scope for innovation

Capabilities of R-software

- Can read any type of data file, viz, Excel, csv, txt, MS-Access, SAS, SPSS, etc.
- Can handle Big Data
- Data preparation and manipulation
- Basic as well as advanced statistical analysis
- Text and Image analytics
- Visualisation is a treat
- Application in GIS
- Greatly helpful for machine learning techniques
- Creating Web applications

Basic Codes for Reading a File and Data Manipulation as Given Below

```
# Setting working directory
setwd ("C:\\Users\\veda\\Downloads")

# Reading file
tomato = read.csv ("Data_Tomato_MultipleMarkets_Tomato.csv")
View (tomato) # Viewing the file in R
names (tomato) # Getting the column or variable names
dim (tomato) # Dimension of the file i.e., number of rows and columns
str (tomato) # Structure of the data
table (tomato$Market) # Frequency table
head (tomato) # Few rows of data
```

```
#Aggregating the Arrivals on Market
market_arrivals = aggregate (tomato $ Arrival, by = list (tomato $ Market),
FUN=sum)

# install.packages ("dplyr") # Installing the dplyr package
library (dplyr) # Invoking the library

# Aggregating the Arrivals on Market
market_arrivals = tomato %>% group_by (Market) %>%
summarize (Qty = sum (Arrival)

# Sorting arrivals
market_arrivals[order (-market_arrivals$Qty) ,]

# Obtaining boxplot
barplot (market_arrivals$Qty)
boxplot (market_arrivals$Qty)

# boxplot (tomato$Modal)
names (tomato)

#Subsetting the data only for KOLAR market
kolar = tomato [tomato$Market == 'KOLAR',]
View (kolar)
boxplot (kolar$Modal)

# Ploting the line graph
plot (kolar$Modal, type='l', col='red', main='Tomato Prices in Kolar',
                              ylab='Prices', xlab='Time')

# Obtaining basic statistics
mean (kolar$Modal)
mean (kolar$Arrival)
sd (kolar$Modal)
min (kolar$Arrival)
max (kolar$Arrival)
median (kolar$Arrival)
sum (kolar$Arrival)
```

4

Statistical Design and Experimental Planning

Dhandapani A.

Experimental Designs

Experimental Designs are used widely to compare more than 2 groups, commonly called as treatments. It is very common in agriculture to conduct several experiments, mostly in fields to test various varieties, different levels of inputs, etc. When one is interested in crop modelling, it is necessary to identify the values of several constants, such as input efficiency, growth periods, growth rates, etc. Instead of guessing these values, it would be more appropriate to conduct experiments to derive these values empirically, which can improve the modelling efforts. The principles of experimentation were given by Sir R.A. Fisher during 1920's, which are still relevant and found in every scientific research namely, (i) Randomization, (ii) Replication and (iii) Local Control. The following sections introduce some of the commonly used terms in Experimental Designs:

Factor: Variables which are set/considered by the experimenters in any experiment. Examples of factors are: Nitrogen, Varieties etc.

Level: The various values of a particular factor considered in the experiment. Examples of levels are: Rates of nitrogen applied, different varieties considered in the experiments.

Treatment: The levels of one or more variables (Factors) set/considered by the experimenters.

Note that in many experiments, the number of variables (i.e. factors) considered may be one only.

Experimental Unit: The unit on which the treatments are applied. For example, in an agronomy experiment, the particular piece of land could be the experimental unit. Other examples of experimental units are: pots, medium, growth chambers etc.

Observational Unit: The units on which the observations are taken. For example, while observing the yield in an experiment, the plot is the observational unit.

The observational units are dependent on the characteristics to be measured. Consider an experiment in which various varieties are to be compared for various characteristics such as yield, plant height, seed weight etc. The experimental units in this experiment are the plots. The observational units for yield are plots whereas for plant height, randomly selected plants within the plots are observational units. It should be kept in mind, while deciding about the experimental layout, what are experimental units and observational units for all the characteristics to be measured from the experiment. The replications are required for experimental units, hence whenever the experimental units and observational units are different, make sure that the experimental units are replicated and not simply the observational units.

Experimental Layouts

The layout of an experiment depends on the type of experimental units, the type of treatments and the practical convenience. To compare the treatments, the fundamental requirement is that the experimental units are same in all respects. If this assumption is satisfied, then it would be possible to compare the treatments. Many a times, it would not be possible to have experimental units which are homogenous in all aspects, particularly, when the number of treatments is more. In such cases, it is advised to form blocks of experimental units in such a way that the units within the blocks are same.

Depending on the experimental units, the most commonly used experimental layouts are: (i) Completely Randomized Designs (CRD), wherein all experimental units are assumed to be homogenous and (ii) Randomized Block Designs (RBD), where it is assumed that the experimental units within a block are homogenous.

In some cases, the treatments are combinations of levels of more than one factor. For example, the responses of varieties exposed under different stress conditions are to be evaluated. In that case, each treatment is a variety to be tried in one of the stress condition. Such experiments are commonly known as Factorial Experiments. Factorial experiments are conducted in CRD depending on experimental units. Layouts of commonly used designs are shown at the end.

In a randomized block design, all the treatments appear once in every block. However, in some experimental situations, it may not be possible to accommodate several treatments within a block without compromising the homogeneity conditions. In such cases, it would be advisable to use Incomplete Block Designs, wherein, only a select set of treatments appear within a block. Several incomplete block designs are available in the literature such as Balanced Incomplete Block Design, Alpha Designs etc. Such designs can provide more accurate comparisons as compared to a complete block design.

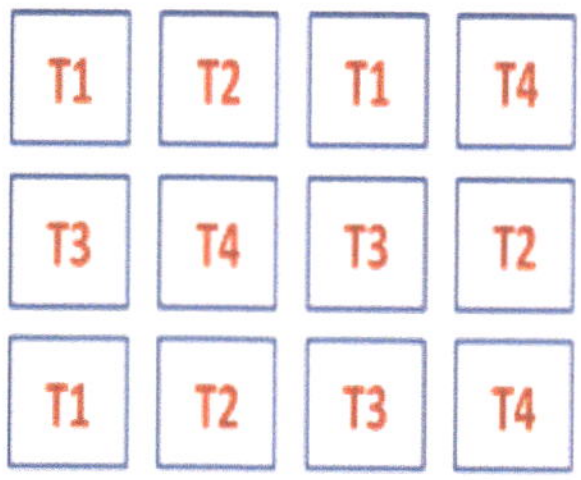

Completely Randomized Design with
4 treatments with 3 replications
(Note: Treatments are randomly allocated)

Randomized Block Design with
6 treatments and 3 Replications

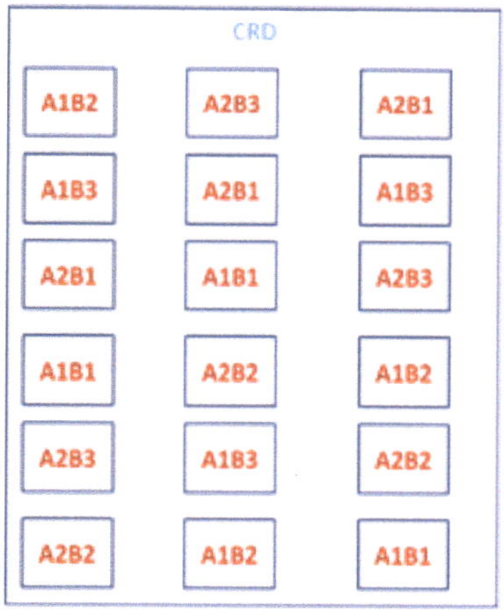

Factorial experiments with one factor at
2 level and another at 3 level conducted
in CRD with 3 Replications

Factorial experiments with one factor at
2 level and another at 3 level conducted
in RBD with 3 Replications

5

Descriptive Statistics, Graphical Analysis, Info Graphics Growth Rate Analysis and Data Mining Techniques in Fisheries

Neha Wajahat Qureshi

Introduction

Importance of basic statistics in everyday life cannot be stressed enough. Inclusion of basic statistics in high school curriculum is uncommon but proves highly beneficial to supplement certain careers in future. This field may be the most widely used of all and most people don't even know they are using it. Many of today's brightest business minds use this study to create charts and graph that allow them to stress a certain aspect of the business they are dealing with. Market research agencies, News rooms, marketing divisions of large corporations, NGOs, accounting firms etc. are full of people who specialize in basic statistics.

Statistics is concerned with the describing, interpreting and analyzing data. It is, therefore, an essential element in any improvement process. Statistics is often categorized into descriptive and inferential statistics. It uses analytical methods which provide the math to model and predict variation. It uses graphical methods to help making numbers visible for communication purposes.

Descriptive Statistics

- Descriptive statistics summarizes or describes characteristics of a data set.
- They help to make sense of large numbers of individual responses, to communicate the essence of those responses to others
- Helps exploring and making conclusions about the data in order to make rational decisions.
- Includes calculating things such as the average of the data, its spread and the shape it produces.

- Are displayed as tables, charts, percentages, frequency, distributions and reported as measures of central tendency
- Determine if the sample is normally distributed (bell curve) as most statistical tests require the sample to have normal distribution
- Determine if the sample can be compared to the larger population

Difference between Descriptive and Inferential Statistics

Nowadays, statistics is playing a major role in the field of research; that helps in the collection, analysis and presentation of data in a measurable form. It is quite hard to identify, whether the research relies on descriptive statistics or inferential statistics, as people usually, lacks knowledge about these two branches of statistics. As the name suggests, descriptive statistics is one which describes the population. On the other end, Inferential statistics is used to make the generalisation about the population based on the samples. So, there is a big difference between descriptive and inferential statistics, i.e., what we do with our data (Table 1).

Table 1: Key differences between Descriptive and Inferential Statistics

Basis for comparison	Descriptive statistics	Inferential statistics
Meaning	Descriptive Statistics is that branch of statistics which is concerned with describing the population under study.	Inferential Statistics is a type of statistics that focuses on drawing conclusions about the population, on the basis of sample analysis and observation.
What it does?	Organize, analyze and present data in a meaningful way.	Compares, test and predicts data.
Form of final result	Charts, Graphs and Tables	Probability.
Usage	To describe a situation.	To explain the chances of occurrence of an event.
Function	It explains the data, which is already known, to summarize sample.	It attempts to reach the conclusion to learn about the population that extends beyond the data available.

Source: https://keydifferences.com/difference-between-descriptive-and-inferential-statistics.html

Types of Descriptive Statistics

There are 4 broad categories under which descriptive statistics of a data set are explained *viz.*

1. Measures of Central Tendency/Position
2. Measures of Variability/Spread

3. Measures of Relative Positions
4. Relationships/ Exploratory Data Analysis

If assignable causes of variation are affecting the process, we will see changes in Position, Spread, Shape or any combination of the three.

1. Measures of Central Tendency/Position

- Measures of central tendency describe the centre of a data set.
- Central tendency refers to where the data is centred.
- Despite the common use of average, there are different statistics by which we can describe the average of a data set: Mean, Median and Mode.

a. Mean

- The total of all the values divided by the size of the data set.
- It is the most commonly used statistic of position.
- It is easy to understand and calculate.
- It works well when the distribution is symmetric and there are no outliers.
- The mean of a sample is denoted by 'x-bar'.
- The mean of a population is denoted by 'μ'.
- Crucial for inferential statistics
- Sensitive to outliers
- Formula for calculating means is $\bar{X} = \frac{\sum_{i=1}^{n} x_i}{n}$

b. Median

- The middle value where exactly half of the data values are above it and half are below it.
- A useful statistic due to its robustness.
- It can reduce the effect of outliers as it is resistant to effects of outliers
- Often used when the data is non-symmetrical.
- Ensure that the values are ordered before calculation.
- With an even number of values, the median is the mean of the two middle values.
- Less useful for inferential purposes and hence less widely used.

c. Mode

- The value that occurs the most often in a data set.
- It is rarely used as a central tendency measure

- It is more useful to distinguish between unimodal and multimodal distributions when data has more than one peak.
- In skewed data, the mean and median lie further toward the skew than the mode.

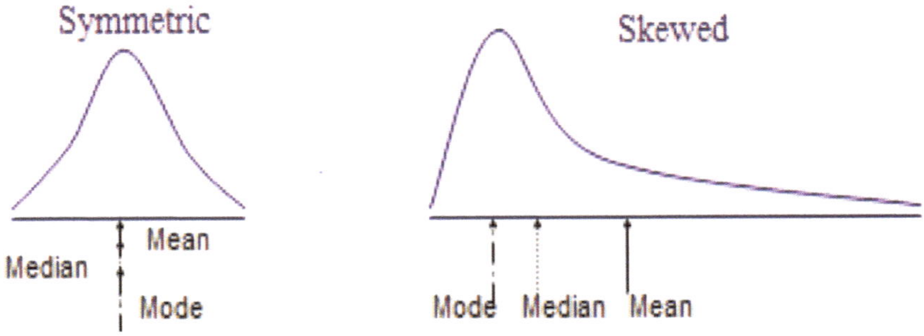

d. Mid-Range

- The mid-range is defined as the sum of the lowest and highest values in the data set, divided by 2.
- The symbol MR is used for the midrange
- Formula is $\frac{Lowest\ value + Highest\ value}{2}$

  Example:

  2,3 6,8,4,1

  MR= 1+8/2= 4.5

2. Measures of Variability / Dispersion / Spread

- Measures of variability or spread describe the dispersion of data within the set.
- Measures of dispersion summarize the degree of clustering/spread of cases, esp. with respect to central tendency.
- The Spread refers to how the data deviates from the position measure.
- It gives an indication of the amount of variation in the process.
- An important indicator of quality.
- Used to control process variability and improve quality.
- Like the mean, these statistics are only meaningful for interval or ratio-level measures
- There are different statistics by which we can describe the spread of a data set:

- Range
- Standard deviation
- Variance
- Coefficient of Variation
- Inter-quartile range

a. Range

- Difference between the highest and the lowest values.
- The simplest measure of variability.
- Often denoted by 'R'.
- It is good enough in many practical cases.
- It does not make full use of the available data.
- It can be misleading when the data is skewed or in the presence of outliers.
- Just one outlier will increase the range dramatically.

b. Standard Deviation

- A summary statistic of how much scores vary from the mean.
- The average distance of the data points from its own mean.
- A low standard deviation indicates that the data points are clustered around the mean.
- A large standard deviation indicates that they are widely scattered around the mean.
- The standard deviation of a sample is denoted by 's'.
- The standard deviation of a population is denoted by "σ".
- It is a more robust measure of variability.
- Standard deviation is computed as follows:
- $S = \sqrt{\frac{\sum_{i=1}^{n}(x_i - \overline{x})^2}{n-1}}$ where,

  s = standard deviation

  x = mean

  x = values of the data set

  n = size of the data set

It is also explained as Square root of the Variance.

- Expressed in the original units of measurement.
- Represents the average amount of dispersion in a sample.
- Used in a number of inferential statistics.

c. Variance

Statisticians use variance to see how individual numbers relate to each other within a data set. To investors, variability is volatility, and volatility is a measure of risk. Therefore, the variance statistic can help determine the risk an investor assumes when purchasing a specific security.

- Average of squared distances of individual points from the mean.
- High variance means that most scores are far away from the mean.
- Low variance indicates that most scores cluster tightly about the mean.
- The amount that one score differs from the mean is called its deviation score (deviate).
- The sum of all deviation scores in a sample is called the *sum of squares.*
- One drawback to variance is that it gives added weight to outliers, the numbers that are far from the mean.
- Squaring these numbers can skew the data.

d. Coefficient of Variation

- The coefficient of variation, denoted by CVar, is the standard deviation divided by the mean.
- It is measure of the dispersion of data points in a data series around the mean.
- Also called Relative Standard Deviation.
- Useful statistic for comparing the degree of variation from one data series to another, even if the means are drastically different from one another.
- The lower the ratio of the standard deviation to mean return, the better risk-return trade-off.
- The result is expressed as a percentage.

$$\text{Cvar}= \frac{s}{\bar{x}} * 100 \text{ for sample}$$

$$\text{Cvar}= \frac{\sigma}{\mu} * 100 \text{ for population}$$

e. Inter-quartile Range

- The Inter Quartile Range is also used to measure variability based on dividing a data set into quartiles.
- Quartiles divide an ordered data set into 4 parts.
- The values that divide each part are called the first, second, and third quartiles
- Each contains 25% of the data.
- The inter quartile range contains the middle 50% of the data (i.e. Q3-Q1).
- It is often used when the data is not normally distributed.
- Interquartile range IQR = UQ – LQ

3. Measures of Shape

- Data can be plotted into a histogram to have a general idea of its shape, or distribution.
- The shape can reveal a lot of information about the data.
- Data will always follow some know distribution.
- It may be symmetrical or non-symmetrical.
- In a symmetrical distribution, the two sides of the distribution are a mirror image of each other.
- Examples of symmetrical distributions include:
    - Uniform
    - Normal
    - Camel-back
    - Bow-tie shaped

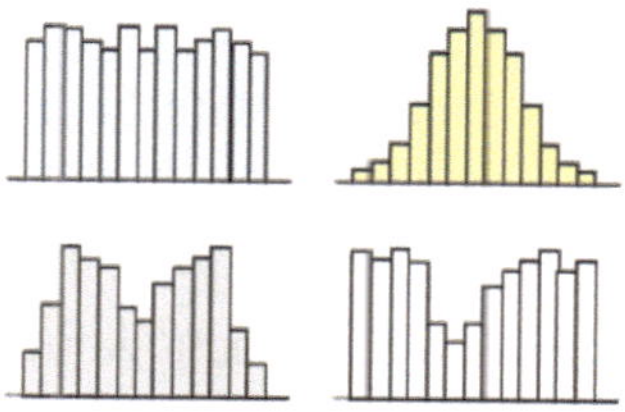

- The shape helps identifying which descriptive statistic is more appropriate to use in a given situation.
- If the data is symmetrical, then we may use the mean or median to measure the central tendency as they are almost equal.
- If the data is skewed, then the median will be a more appropriate to measure the central tendency.
- Two common statistics that measure the shape of the data:
    - Skewness
    - Kurtosis

a. Skewness

- Measures look at how lopsided distributions are—how far from the ideal of the normal curve they are.
- When the median and the mean are different, the distribution is skewed. The greater the difference, the greater the skew.
- Distributions that trail away to the left are negatively skewed and those that trail away to the right are positively skewed
- If the skewness is extreme, the researcher should either transform the data to make them better resemble a normal curve or else use a different set of statistics—non-parametric statistics
- A skewness value of zero indicates perfect symmetry.
- A negative value implies left-skewed data.
- A positive value implies right-skewed data.

$$S_{k=3}\frac{(\bar{Y}-\tilde{Y})}{s}$$

Where: where Y~ is the sample median.

Y¯ is the mean, **s** is the standard deviation

b. Kurtosis

- Measures the degree of flatness (or peakness) of the shape.
- Measure of whether the data are heavy-tailed or light-tailed relative to a normal distribution.
- Data sets with high kurtosis tend to have heavy tails, or outliers. Data sets with low kurtosis tend to have light tails, or lack of outliers.
- The histogram is an effective graph (skewness and kurtosis)
- When the data values are clustered around the middle (peaked)
- When the data values are spread around more evenly (flattened).
- Normal Distribution value of Kurtosis is 3

$$\text{Kurtosis} = \frac{\sum_{i=1}^{N}(Y_i-\bar{Y})^4 \,/\, N}{s^4} - 3$$

Where; Y¯ is the mean, **s** is the standard deviation, and *N* is the number of data points.

4. Measure of Relationships/ Exploratory Data Analysis (EDA)

- Exploratory data analysis (EDA) is an approach to analyze the data sets to summarize their main characteristics, often with visual methods.

- Primarily EDA is for seeing what the data can tell us beyond the formal modelling or hypothesis testing task.
- The purpose of exploratory data analysis is to:
  - Check for missing data and other mistakes.
  - Gain maximum insight into the data set and its underlying structure.
  - Uncover a parsimonious model, one which explains the data with a minimum number of predictor variables.

Graphical Analysis

- The graphs we choose depends on:
  - The type of data we have.
  - The objective we are trying to achieve.

There are graphs for continuous data and graphs for count and attribute data. Remember that we need to perform additional statistical analysis before drawing any conclusion.

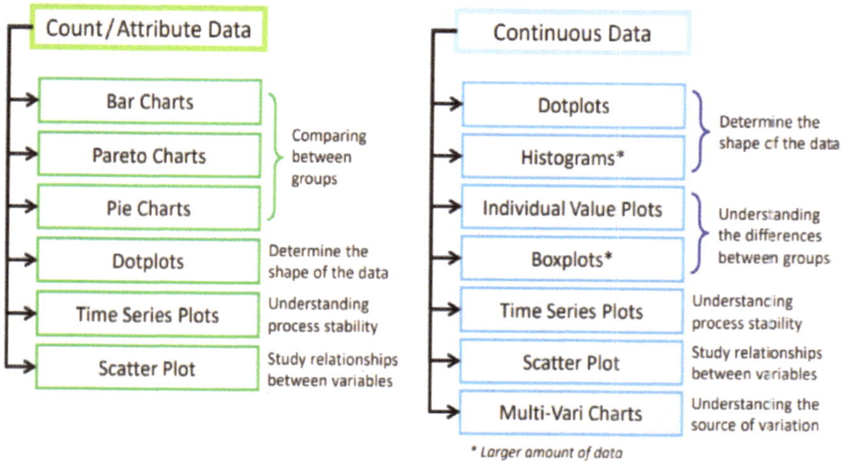

Source: Continuous Improvement Toolkit

Statistics may be:

- **Graphical:** Makes the numbers visible.
- **Inferential:** Makes inferences about population from sample data.
- **Analytical:** Uses math to model and predict variation.
- **Descriptive:** Describes characteristics of the data (location and spread)
  - Graphs truly show that a picture is worth thousands of words.

- A long list of data is usually not practical for conveying information about a process.
- One of the best ways to analyze any process is to plot the data.

Importance of Graphs

- Allows learning about the nature of the process.
- Enables clarity of communication.
- Helps understanding sources of variation in the data.
- Provides focus for further analysis.
- Different graphs can reveal different characteristics of our data:
  - Central tendency
  - Dispersion
  - Shape for the distribution

Conclusions drawn from graphs may require verification through advanced statistical techniques such as significance testing and experimentation.

Types of Graphs

There are numerous types of graphs available and depending upon the type of data and need of the researcher suitable graphs can be selected. Some important types of graphs often used in data analysis in social science are discussed below:

1. Line Charts

- One of the simplest form of charts.
- Useful for showing trends in quality, cost or other process performance measures.
- They represent the data by connecting the data points by straight lines to highlight trends in the data.
- A standard line may also be drawn to verify actual performance against identified targets.
- Time series plots, run charts, SPC charts and radar charts are all line charts.

2. Time Series Plots

- Line charts used to evaluate behavior in data over a time interval.
- They can be used to determine if a process is stable by visually spotting trends, patterns or shift in the data.

- If any of these are observed, then we can say that the process is probably unstable.
- It requires the data to be in the order which actually happened.

Time Series Analysis is the analysis of the plotted data in order to get meaningful information.

Different behaviours of the data can be observed such as:

- Upward and downward trends.
- Shifts in the mean.
- Changes in the amount of variation.
- Patterns and cycles

Time Series Forecasting is the use of a model to predict future values based on observed values.

3. Pie Charts

- Circular charts that make it easy to compare proportions.
- Widely used in the business and media world for their simplicity and ease of interpretation.
- They represent each category as a slice of the pie.
- They display the proportion of each category relative to the whole data set.

Example: Change in market percentage of seafood exports quantity-wise and value-wise.

4. Doughnut Charts

- A Doughnut Chart is a variation of the pie chart with a blank center.
- It allows for additional information to be included about the data.
- Pie and doughnut charts work well with few categories.
- They are suitable for presenting data for around seven groups or fewer.
- Examples : Species composition of a reservoir/ Ingredients used in a value added product

5. Bar Charts

- Used to display frequencies of attribute data.
- They focus on the absolute value of the data.
- The bars on the chart are presented horizontally or vertically.
- When a bar chart presents the categories in descending order of frequency, this is called a Pareto Chart.

- Grouped Bar Charts display bars clustered in groups.
- Staked Bar Charts stack bars of each group on top of each other to show the cumulative effect.
- A 100% Staked Bar Chart is used for demonstrating the difference in proportion between categories.

6. Dot Plots

- A dot plot is a graphical representation of data using dots plotted on a simple scale.
- A form of frequency distribution.
- It is suitable for displaying small to moderate data sets.
- The X-axis is divided into many small intervals called bins.
- The data values falling within each bin are represented by dots (one or more dots per data point).
- The end result is a set of vertical lines of dots.
- It is generally used when the data is discrete.
- It can also be used to present continuous data.
- It shows where the data are clustered, where the gaps are located and can help identify outliers.
- Dotplots are also useful for comparing distributions in terms of their shape, location and spread.

7. Histogram

- Shows center, variability, skewness, modality, Outliers, or strange patterns.
- Bin width and position matter
- Beware of real zeros

Issues with Histograms

- For small data sets, histograms can be misleading.
  Small changes in the data, bins, or anchor can deceive.
- For large data sets, histograms can be quite effective at illustrating general properties of the distribution.
- Histograms effectively only work with 1 variable at a time
  - But 'small multiples' can be effective

8. Individual Plots

- Graphs that are useful to give an overall picture of the individual values that make up a data set.
- Often used for comparing distributions that have small number of data.
- They enable to see all the values of a dataset even if there are similar data points.
- They give an idea of the distribution shapes and whether outliers are present.

9. Radar Charts

- Used to display and compare multiple data sets over a range of characteristics or over a specific period of time.
- It comes in the form of a two-dimensional chart.
- It has a radial axis and an angular axis.
- After plotting the data, a point close to the center indicates a low value and a point near the edge indicates a high value.
- A line is normally drawn connecting the data values for each data set.

  Example: plot indices like livelihood index, perception index, vulnerability index etc.

10. Multi-variate Charts

- Variation in the data may come from multiple sources.
- A Multi-vari chart is a graphical tool that allows to visually show where the major variation is coming from.
- Multiple variables are plotted together on a single chart.
- Often used when studying the variation within:
  - A subgroup
  - Between subgroups
  - Over time

11. Scatter Plots

- Many problems require the estimation of the relationship between two or more variables.
- Scatter plots are used to study the relationship between two variables.
- They are used to determine what happens to one variable when another variable changes value.

  Example: Plotting CPUE against no. of fishermen

12. Probability Plots

- Graphical techniques that provide a more decisive approach for determining if your data follows a particular distribution.
- Constructed in a way that the points will fall in a straight line if they fit the distribution in question.
- They are an improvement from just assessing visually.

Example P-P plots and Q-Q plots

13. Box Plots

- A **box plot** (also known as **box and whisker plot**) is a type of chart often used in explanatory data analysis to visually show the distribution of numerical data and skewness through displaying the data quartiles (or percentiles) and averages.

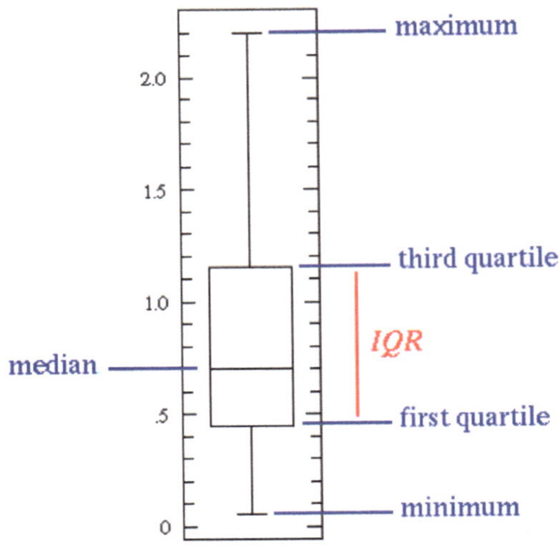

R-Codes and Libraries for Graphical Analysis

barplot (height = data $ value, names = data $ name)	**ggplot** (data, **aes** (x= **x-value**, y = y-value)) + **geom_boxplot** (fill = "slateblue", alpha = 0.2) + **xlab** ("cyl")	**ggplot** (**aes** (x=-price)) +**geom_density** (fill= "#69b3a2", color= "#e9ecef", alpha=0.8)	**ggplot**(data, **aes** (fill= condition, y= value, x= value)) + **geom_bar** (position= "dodge", stat= "identity")	**ggplot (data, ae s (x= value))** + **geom_histogram ()**

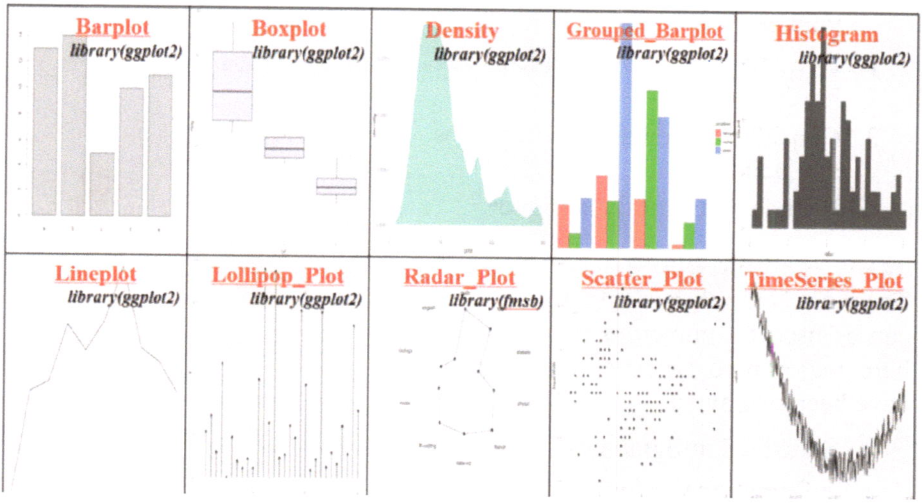

ggplot(data, **aes**(x=xValue, y=yValue)) + **geom_line()**	**ggplot**(data, **aes**(x=x, y=y)) + **geom_point()** + **geom_segment(** **aes**(x=x, xend=x, y=0, yend=y))	**library** (fmsb) **radarchart** (data)	**ggplot** (data, **aes** (x=x-value, y=y-value)) + **geom_point()**	**ggplot** (data, **aes** (x=day, y=value)) + **geom_line()**+ **xlab**("")

Source: https://www.r-graph-gallery.com/index.html

Info Graphics

- An infographic is a collection of imagery, charts, and minimal text that gives an easy-to-understand overview of a topic.
- Infographics use striking, engaging visuals to communicate information quickly and clearly.

Why are Infographics used?

- Infographics are great for making complex information easy to digest.
- Provide a quick overview of a topic hard to explain.

- Explain a complex process.
- Display research findings or survey data.
- Summarize a long blog post or report.
- Compare and contrast multiple options.
- Raise awareness about an issue or cause.

Steps to create Inforgraphics

1. Determine the key takeaways of your content.
2. Determine the title, headers, sub headers and facts.
3. Consider the length of paragraphs and points.
4. Include notes for the designer.

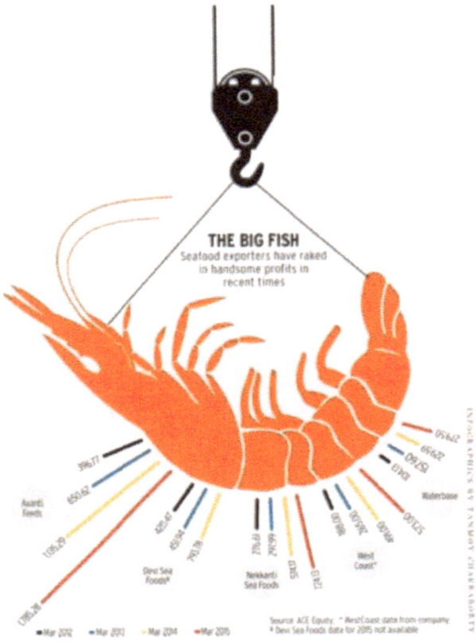

Types of Infographics

There are N number of infographics that can be used for numerous purposes, but here major nine types of infographics have been explained.

- Statistical infographics
- Informational infographics
- Timeline infographics
- Process infographics
- Geographic infographics
- Comparison infographics
- Hierarchical infographics
- List infographics
- Resume infographics

What makes an Infographic Effective

- Brain seeks patterns in visual information to help us make sense of the world.
- We can use this idea to structure our information visually and create patterns that will enhance the message that we're trying to communicate.
- Use lines, borders, and shapes to group related information.
- Use one contrasting color to guide your readers' attention.

- Pick one color that contrasts with all of the other colors in the graphic, and use it to make the most important information stand out.
- Create a text hierarchy with three different font styles.
- Use images, icons, and illustrations to make key takeaways memorable.

Useful links below for assessing few Infographic templates

https://venngage.com/templates/

https://www.canva.com/en_in/create/infographics/

https://piktochart.com/formats/infographics/

https://www.easel.ly/home

https://infograpia.com/collections/infographics

Growth Rate Analysis

What are growth rates?

Growth rates refer to the percentage change of a specific variable within a specific time period.

For investors, growth rates typically represent the CAGR of growth of a company's revenues, earnings, dividends or even macro concepts, such as (GDP) and retail sales. Growth rates can be beneficial in assessing a company's performance and to predict the future performance.

Growth rates are used to express the annual change in a variable as a percentage, such as revenues or investments. It is the amount in which the value of an investment, asset, portfolio or business increases over a specific period. The growth rate provides us with important information about the value of an asset or investment as it helps us understand how that asset or investment grows, changes and performs over time. This information can help us predict the future revenue of a specific asset or investment.

Growth rate can be used to represent

- The percent change in the overall population from one year to another
- The performance of a business and its expected future growth and
- The performance and expected future growth of other assets and investments.

Methods of Calculating Growth Rates

There are broadly three methods of calculating growth rates

- Straight-line percent change method
- Midpoint method
- Calculating the average growth rate over time

a. Straight-line Percent Change Method

- Most common formula used to calculate basic, simple growth rates.
- This growth rate calculation method works best when calculating simple growth rates that we don't need to compare to other results.
- This formula may not produce uniform results for negative changes in growth.
- This limitation of the method is known as the end-point problem.
- (Growth Rate = Absolute Change / Original Value)

  (Absolute Change = New Value - Original Value)

  (Growth Rate = Absolute Change / Original Value)

  (Percent Change = Growth Rate x 100)

b. Mid-Point Method

- Used to calculate growth rate when we need to use the information to make comparisons.
- Benefit - It provides uniform results regardless of the direction of change and avoids the end-point problem
- The midpoint method overcomes this challenge by using the average of the original and ending values as the denominator.
- (Growth Rate = Absolute Change / Average Value)
  - (Absolute Change = New Value - Original Value)
  - (Average Value = Sum of Values / 2)
  - (Percent Change = Growth Rate x 100)

c. Average Growth Rate Overtime

- When we are comparing the change in growth for several years, it is best to use this calculation method to find the average growth rate per year.
- [Growth Rate = (Present Value / Past Value) 1/N - 1]

  (Present Value - Past Value)

  N= No. of years

  (Percent Change = Growth Rate x 100)

Types of Growth Rates

- Average annual growth rate (AAGR)
- Straight-line Growth Rate (i.e. % change of final and beginning figure)

- Compound Annual Growth Rate (CAGR)
- Exponential Growth Rate

1. Average Annual Growth Rate (AAGR)

- Average increase in the value of an individual investment, portfolio, asset, or cash stream over the period of a year.
- It is calculated by taking the arithmetic mean of a series of growth rates.
- The average annual growth rate can be calculated for any investment, but it will not include any measure of the investment's overall risk, as measured by its price volatility.
- This ratio helps to figure out how much average return is received over several periods of time.
- AAGR is a linear measure that does not account for the effects of compounding.

$$AAGR = \frac{GR_A + GR_B + ...GR_n}{N}$$

where,

GR_A = Growth rate in period A

GR_B = Growth rate in period B

GR_n = Growth rate in period n

N = Number of payments

- The AAGR measures the average rate of return or growth over a series of equally spaced time periods. As an example, assume an investment has the following values over the course of four years:
    - Beginning value = \$100,000
    - End of year 1 value = \$120,000
    - End of year 2 value = \$135,000
    - End of year 3 value = \$160,000
    - End of year 4 value = \$200,000

The formula to determine the percentage growth for each year is:

Simple percentage growth or return = (ending value/ Beginning value) -1

Thus, the growth rates for each of the years are as follows:

Year 1 growth = \$120,000 / \$100,000 - 1 = 20%

Year 2 growth = \$135,000 / \$120,000 - 1 = 12.5%

Year 3 growth = \$160,000 / \$135,000 - 1 = 18.5%

Year 4 growth = \$200,000 / \$160,000 - 1 = 25%

$$\text{AAGR} = \frac{20\% + 12.5\% + 18.5\% + 25\%}{4} = 19\%$$

- In the financial and accounting settings, typically the beginning and ending prices are used, but some analysts may prefer to use average prices when calculating the AAGR depending on what is being analyzed.

Screenshots of excel sheet depiciting formulas for calculating AAGR

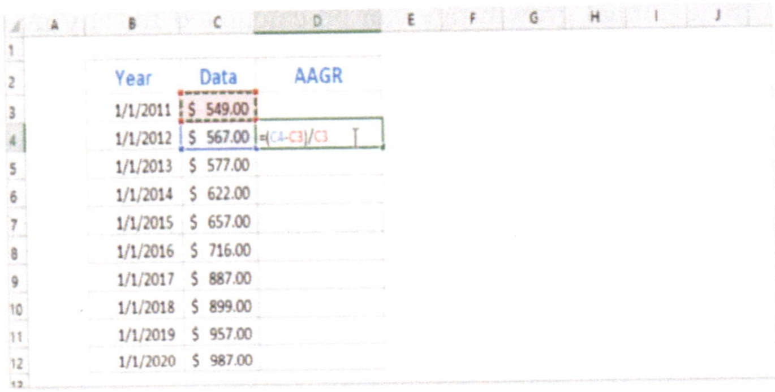

Year	Data	AAGR
1/1/2011	$ 549.00	
1/1/2012	$ 567.00	=(C4-C3)/C3
1/1/2013	$ 577.00	
1/1/2014	$ 622.00	
1/1/2015	$ 657.00	
1/1/2016	$ 716.00	
1/1/2017	$ 887.00	
1/1/2018	$ 899.00	
1/1/2019	$ 957.00	
1/1/2020	$ 987.00	

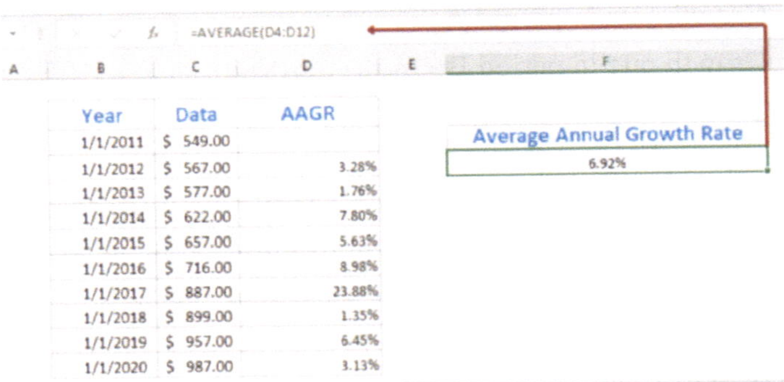

Year	Data	AAGR
1/1/2011	$ 549.00	
1/1/2012	$ 567.00	3.28%
1/1/2013	$ 577.00	1.76%
1/1/2014	$ 622.00	7.80%
1/1/2015	$ 657.00	5.63%
1/1/2016	$ 716.00	8.98%
1/1/2017	$ 887.00	23.88%
1/1/2018	$ 899.00	1.35%
1/1/2019	$ 957.00	6.45%
1/1/2020	$ 987.00	3.13%

2. Compound Annual Growth Rate

- Compound annual growth rate (CAGR) is a method used to calculate annual grow rate from time series.
- The result of CAGR is interpreted as the smoothed annualized growth rate achieved during the considered time horizon.
- It therefore represents the rate at which the variable would have grown if the rate of growth was constant during the considered period.
- The CAGR smooths out an investment's returns or diminishes the effect of volatility of periodic returns.

$$\text{CAGR}= \left(\frac{\textit{End value}}{\textit{Beginning value}}\right)^{1/n} -1$$

n = no. of years

Screenshots of excel formulas for calculating CAGR

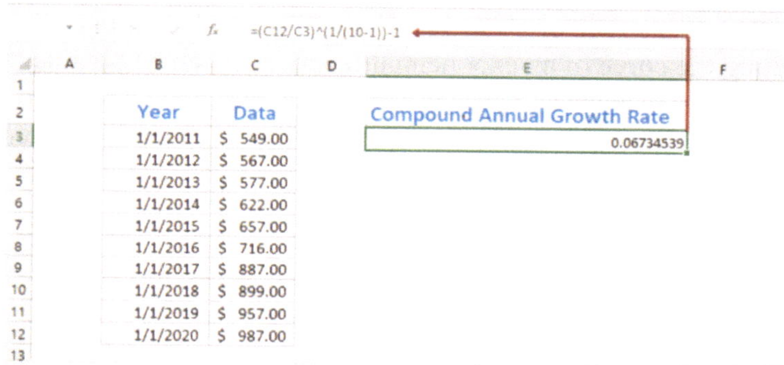

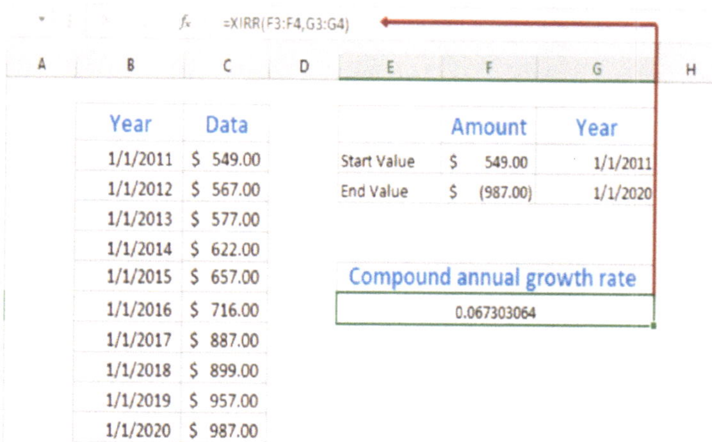

Applications of CAGR

- Calculating and communicating the average returns of investment funds.
- Demonstrating and comparing the performance of investment advisors.
- Comparing the historical returns of stocks with bonds or with a savings account.
- Forecasting future values based on the CAGR of a data series (find future values by multiplying the last datum of the series by (1 + CAGR) as many times as years required). As every forecasting method, this method has a calculation error associated.

- Analyzing and communicating the behavior, over a series of years, of different business measures such as sales, market share, costs, customer satisfaction, and performance.

3. Straight Line Percent Method (discussed above)

4. Exponential Growth Rate

Exponential growth is a pattern of data that shows greater increases with passing time, creating the curve of an exponential function. It can be obtained by plotting the time series in excel and getting the exponential trend line fitted in along with R-square values and equation on the chart as shown below. The exponential growth rates can be obtained accordingly.

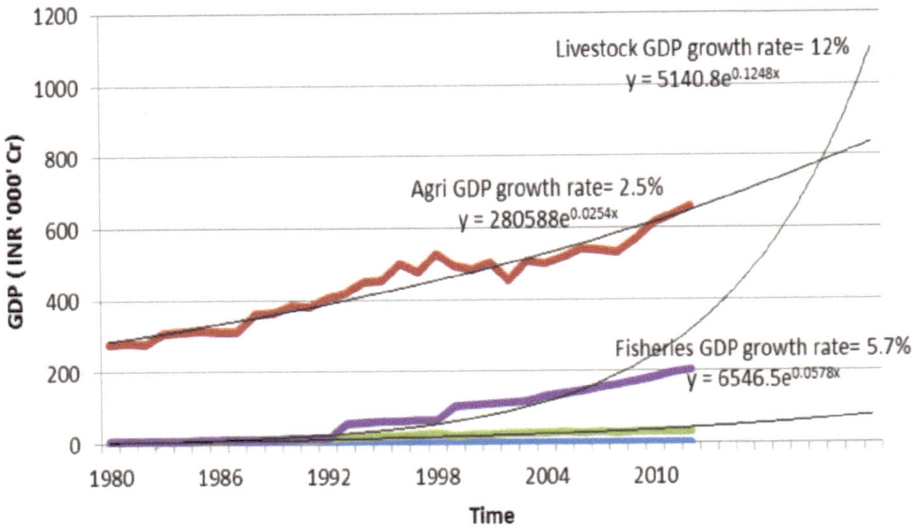

Figure: Showing Exponential growth rate of Agricultural, Livestock and Fisheries GDP (Qureshi *et al.*, 2015)

Data Mining Techniques

Data mining is the semi-automatic discovery of patterns, associations, changes, anomalies, and statistically significant structures and events in data. Traditional data analysis is an assumption driven in the sense that a hypothesis is formed and validated against the data. Data mining, in contrast, is data driven in the sense that patterns are automatically extracted from data.

- Data Mining is an important analytic process designed to explore data.
- Process of extraction of useful information and patterns from huge data.

- Also called as knowledge discovery process, knowledge mining from data, knowledge extraction or data /pattern analysis.
- It is incredibly important to spend appropriate time in "massaging" the data, extracting important information before moving forward with the modelling.
- The tasks of data mining are two-fold:
    - Create predictive power—using features to predict unknown or future values of the same or other feature—
    - Create a descriptive power—find interesting, human-interpretable patterns that describe the data

Steps Involved in Data Mining

Data mining is a logical process that is used to search through large amount of data in order to find useful data. The goal of this technique is to find patterns that were previously unknown. Once these patterns are found, they can further be used to make certain decisions for development of their businesses.

There are three broad steps:

a. Exploration
b. Pattern Identification
c. Deployment Exploration

Exploration

In the first step of data exploration, data is cleaned and transformed into another form, and important variables and then nature of data based on the problem are determined.

Pattern Identification

Once data is explored, refined and defined for the specific variables, the second step is to form pattern identification. Identify and choose the patterns which make the best prediction.

Deployment

Patterns are deployed for desired outcome.

Data Mining Algorithms and Techniques

Various algorithms and techniques like Classification, Clustering, Regression, Artificial Intelligence, Neural Networks, Association Rules, Decision Trees, Genetic Algorithm, Nearest Neighbor method etc., are used for knowledge discovery from databases.

1. Classification

Classification is the most commonly applied data mining technique, which employs a set of pre-classified examples to develop a model that can classify the population of records at large. Fraud detection and credit risk applications are particularly well suited to this type of analysis. This approach frequently employs decision tree or neural network-based classification algorithms. The data classification process involves learning and classification. In learning, the training data are analyzed by classification algorithm. In classification test, data are used to estimate the accuracy of the classification rules. If the accuracy is acceptable, the rules can be applied to the new data tuples. For a fraud detection application, this would include complete records of both fraudulent and valid activities determined on a record-by-record basis. The classifier-training algorithm uses these pre-classified examples to determine the set of parameters required for proper discrimination. The algorithm then encodes these parameters into a model called a classifier. Types of classification models:

- Classification by decision tree induction
- Bayesian Classification
- Neural Networks
- Support Vector Machines (SVM)
- Classification based on associations

Clustering

Clustering can be said as identification of similar classes of objects. By using clustering techniques, we can further identify dense and sparse regions in object space and can discover overall distribution pattern and correlations among data attributes. Classification approach can also be used for effective means of distinguishing groups or classes of object, but it becomes costly, so clustering can be used as pre-processing approach for attribute subset selection and classification. For example, to form group of customers based on purchasing patterns, to categories genes with similar functionality.

Types of Clustering Methods

- Partitioning Methods
- Hierarchical Agglomerative (divisive) methods
- Density based methods
- Grid-based methods
- Model-based methods

2. Predication

Regression technique can be adapted for predication. Regression analysis can be used to model the relationship between one or more independent variables and dependent variables. In data mining, independent variables are attributes already known and response variables are what we want to predict. Unfortunately, many real-world problems are not simply prediction. For instance, sales volumes, stock prices, and product failure rates are all very difficult to predict because they may depend on complex interactions of multiple predictor variables. Therefore, more complex techniques (e.g., logistic regression, decision trees, or neural nets) may be necessary to forecast future values. The same model types can often be used for both regression and classification. For example, the CART (Classification and Regression Trees) decision tree algorithm can be used to build both classification trees (to classify categorical response variables) and regression trees (to forecast continuous response variables). Neural networks too can create both classification and regression models.

Types of Regression Methods

- Linear Regression
- Multivariate Linear Regression
- Nonlinear Regression
- Multivariate Nonlinear Regression

3. Association Rule

Association and correlation is usually to find frequent item set findings among large data sets. This type of finding helps businesses to make certain decisions, such as catalogue design, cross marketing and customer shopping behavior analysis. Association Rule algorithms need to be able to generate rules with confidence values less than one. However, the number of possible Association Rules for a given dataset is generally very large and a high proportion of the rules are usually of little (if any) value.

Types of association rule:

- Multi-level association rule
- Multi-dimensional association rule
- Quantitative association rule

References

Bharati M. Ramageri, 2018. Data mining techniques and applications. Indian Journal of Computer Science and Engineering, 1 (4): 301-305.

Quershi, N.W., Achoth, L., Krishnan, M., Prakash, S., Ananthan, P.S., Kumar, N.R. and Pandey, S.K., 2015. Misplaced priorities and lopsided investments: a macroscan of agriculture, livestock and fisheries sectors in India. Agricultural Economics Research Review, 28 (347-2016-17193): 25-34.

Online References

https://www.investopedia.com/

https://www.wikipedia.org/

https://venngage.com/

https://www.r-graph-gallery.com/index.html

https://keydifferences.com/difference-between-descriptive-and-inferential-statistics.html

https://www.slideshare.net/dsaadeddin/graphical-analysis-57152882#:~:text=%EF%81%B1% 20 Often %20 used %20for%20 comparing, Graphical% 20Analysis

6

Cluster Analysis for Researchable Datasets in Fisheries

Ashok Kumar J.

There has been a surge of data in every field of research due to advancements in information technology. Data collection and storage has become much easier and cheaper in recent years. This voluminous data needs to be systematically stored for quick retrieval, process and to find useful patterns. Data mining techniques help in deriving useful patterns from data. These techniques which are also called as machine learning techniques majorly categorized as classification, clustering and association rule mining methods. Classification is a supervised learning method which needs labeled dataset for model building and prediction purposes. Clustering and association rule mining methods do not require labeled dataset and hence they are referred as unsupervised learning methods.

Clustering aims at grouping the objects based on a set of measured variables where similar objects are assigned to one group. For example, let us say, if we have collected information on 50 fish pond parameters (50 variables) and we wish to find the groups in terms of pond management (like well managed to poorly managed), with cluster analysis, we can group the ponds based on pond variables. In the process of clustering, each pond is assigned to one of the groups made. The question is how many natural clusters present in the data as in many cases we would not be getting perfect clusters of disjoint sets. We can achieve better accuracies in cluster formation only by trial and error. Good clustering criteria are to see that there is minimum with-in cluster variations and maximum between cluster variations.

Cluster analysis has several applications in fisheries research. It could be used to group fish stocks from different regions, categorise ponds based on their technical efficiencies, identify the farm clusters with high disease risk, group farmers based on their preferences, agro zoning based on climatic variables and culture practices. Cluster analysis is also applied in fish bioinformatics to group the similar sequences.

There are several clustering algorithms used for clustering the objects. Most common methods include k-means clustering, hierarchical clustering, model based clustering and density based clustering. K-means clustering is a simple iterative method in which one needs to mention the number of clusters (k) to be made for the given dataset. Algorithm of this method starts with selecting the number of clusters and randomly selecting initial cluster averages. It further calculates distance between each object to its average and assigns the objects to its closest mean cluster. The process iterates and every time the objects are adjusted with their distances with new centroids. The distance measure that is most commonly used in this method is the squared Euclidean distance. The k-means clustering method is easy to use and can be a good starting point. Procedure needs to be repeated with different k values for obtaining optimal solutions.

The second most important clustering procedure is hierarchical clustering. Here in this method there is no need to mention the number of clusters before starting the procedure and it outputs a tree with leaf nodes as objects. There are two different algorithms in this method that are widely used. One is agglomerative and the other is divisive. Agglomerative methods follow a bottom up approach i.e., tree building starts with individual objects (leaf nodes) and ends with a single cluster while in a divisive method the entire dataset is considered as one cluster at first and proceeds on dividing them into different groups. In hierarchical clustering, the distance matrix for the set of input variables is calculated at first and the tree is built based on that matrix. The distance measures that are commonly used include Euclidean, maximum, Manhattan and Minkowski and the most common agglomerative methods include ward, average, mcquitty, median and centroid.

Model based clustering assumes data is coming from a distribution that is a mixture of two or more clusters. Unlike other methods mentioned, model based clustering assigns probabilities for each object signifying its belonging to different clusters. Density-based clustering assumes that each cluster in a data space is a contiguous region of high point density, separated from other such clusters by contiguous regions of low point density. Low point density regions are considered as outliers.

Cluster Analysis using R stats

Get data into R

- setwd ('C:\path\to\directory')
- data<- read.csv ('yourfile.csv',h=T,sep=",")

Check the data

- head(data)

Data needs to be scaled to give equal weightage to all the variables considered for clustering

- data1<- scale(data)

K-means clustering with 3 clusters

km1 <- kmeans (data1, 3)

When we print 'Km1' we get the ratio of between sums of squares to the total sum of squares. This ratio value gives indication of optimum number of clusters that can be made with the data. If the increment in K value (number of clusters) does not improve ratio value substantially then that K value is considered as optimum value.

To plot the K-means clusters

- library (cluster)
- clusplot (data1, km1$cluster, color=TRUE, shade=TRUE)

Hierarchical clustering

Making distance matrix

- d1 <- dist (data1, method="euclidean")

Make cluster

- h1 <- hclust (d1, method="ward.D")

Plot cluster

- plot (h1)

Make rectangles of 5 clusters from the tree

- rect.hclust (h1,k=5, border="green")

Model based clustering

- Library (mclust)
- Model<- Mclust (data2)
- Plot (model)

Density based clustering

- Library (dbscan)
- Fit <- dbscan (data1, eps=0.7, min Pts=31)

  min Pts can be one value more than number of variables and eps value is obtaine from following plot
- kNN distplot (data1, k=3)

Several open source tools other than R stats are also available for carrying out the cluster analysis which includes WEKA software (https://www.cs.waikato.ac.nz/ml/weka), Scikit-learn library in Python. Google colab (https://colab.research.google.com/) and Kaggle (https://www.kaggle.com/) offers platforms for directly implementing Python/R code in Jupytor notebooks.

References

Arabie P, De Soete G., 1996, Clustering and classification. World Scientific.

Jain A K., 2008. Data clustering: 50 years beyond k-means. In Joint European Conference on Machine Learning and Knowledge Discovery in Databases 2008 Sep 14: 3-4. Springer, Berlin, Heidelberg.

Online References

https://www.r-project.org/
https://www.cs.waikato.ac.nz/ml/weka
https://www.python.org/
https://scikit-learn.org/
https://colab.research.google.com/
https://www.kaggle.com/

7

Conjoint Analysis in Fisheries Data Sets

V. Geethalakshmi

Product Development and Marketing

Several efforts are being made on the research front for development of novel products from fish which have enhanced shelf life, retain the freshness and nutrients and are safe to consume. During the course of product development, which are in many stages, studies are conducted prior to commercialisation to ensure the acceptance of a product.

In a business, understanding consumer behaviour plays an important role in success. To be in a business for a long time, it is essential to know what the consumer prefers and why. Consumers make the buying decisions based on a number of factors. The purpose behind studying the buying behaviour and consumer preference is to produce and market products which may better meet the needs of a consumer. The emerging fastfood culture among the young and affordable has brought focus on processed food and its demand in the domestic food market in India. Domestically, spending on food and food products constitutes the largest portion of the Indian consumer's spending – more than a 31% share of wallet. Evaluation of consumer preferences before introducing a new product will help the marketer to refine the product for better reach.

Conjoint Analysis

Conjoint analysis is a popular technique used in marketing research to study the features a product that should possess to have a wide consumer reach. Conjoint analysis was initially conceptualised by Luce and Tukey (1964) and further developed by Green and Rao (1971) for marketing research. It employs a decompositional method to estimate the structure of consumer preferences and consumer utility values of different attributes of a product or service. It is a decompositional method that disaggregates the structure of consumer preferences into utility values. The relative importance of a product can also be estimated using this method.

Products possess attributes such as price, color, ingredients, guarantee, environmental impact, predicted reliability, and so on. Consumers typically do not have the option of buying the product that is best in every attribute, particularly when one of those attributes is price. Consumers are forced to make trade-offs as they decide which products to purchase. Consider the decision to purchase a car. Increased size generally means increased safety and comfort. The tradeoff is an increase in cost and environmental impact and a decrease in mileage and manoeuvrability. Conjoint analysis is used to study these trade-offs.

If products are composed of attributes, conjoint analysis determines which combination of attribute levels are most preferred by consumers. Consumers indicate their preferences by ranking a number of different combinations of attribute levels. Conjoint analysis assumes that consumers make purchases by simultaneously considering several attributes of a product. The ability to analyze several attributes at once distinguishes conjoint analysis from traditional market research methods where each attribute is studied separately. Usually, conjoint analysis consists of a main-effects analysis of variance with ordinally scaled dependent variables. Consumer preferences are the dependent variables, and product attributes are the independent variables. The following are some of the questions that can be answered with a conjoint analysis.

- How important is each product attribute to consumers?
- Which existing products do consumers prefer?
- What combination of product attributes do consumers prefer the most?
- How well will my product do in the current market?

Conjoint analysis is based on a main effects analysis-of-variance model. Subjects provide data about their preferences for hypothetical products defined by attribute combinations. Conjoint analysis decomposes the judgment data into components, based on qualitative attributes of the products. A numerical part-worth utility value is computed for each level of each attribute. Large part-worth utilities are assigned to the most preferred levels, and small part-worth utilities are assigned to the least preferred levels. The attributes with the largest part-worth utility range are considered the most important in predicting preference. Conjoint analysis is a statistical model with an error term and a loss function.

Metric conjoint analysis models provide the judgments directly. When all of the attributes are nominal, the metric conjoint analysis is a simple main-effects ANOVA with some specialized output. The attributes are the independent variables, the judgments comprise the dependent variable, and the part-worth utilities are the β's, the parameter estimates from the ANOVA model. The following formula shows a metric conjoint analysis model for three factors:

$$y_{ijk} = \mu + \beta_{1i} + \beta_{2j} + \beta_{3k} + e_{ijk}$$

where $X\beta_{1i} = X\beta_{2j} = X\beta_{3k} = 0$

This model could be used, for example, to investigate preferences for cars that differ on three attributes: mileage, expected reliability, and price. The y_{ijk} term is one subject's stated preference for a car with the ith level of mileage, the jth level of expected reliability, and the kth level of price. The grand meanis μ, and the error is e_{ijk}. The predictedutility for the ijk product is:

$$y\hat{}_{ijk} = \hat{}\mu + \beta\hat{}_{1i} + \beta\hat{}_{2j} + \beta\hat{}_{3k}$$

Nonmetric conjoint analysis finds a monotonic transformation of the preference judgments. The model, which follows directly from conjoint measurement, iteratively fits the ANOVA model until the transformation stabilizes. The R^2 increases during every iteration until convergence, when the change in R square is essentially zero. The following formula shows a nonmetric conjoint analysis model for three factors:

$$\Phi(y_{ijk}) = \mu + \beta_{1i} + \beta_{2j} + \beta_{3k} + e_{ijk}$$

where $\Phi(yijk)$ designates a monotonic transformation of the variable y.

The R^2 for a nonmetric conjoint analysis model is always greater than or equal to the R^2 from a metric analysis of the same data. The smaller R^2 in metric conjoint analysis is not necessarily a disadvantage, since results should be more stable and reproducible with the metric model. Metric conjoint analysis was derived from nonmetric conjoint analysis as a special case. Today, metric conjoint analysis is probably used more often than nonmetric conjoint analysis. In the SAS System, conjoint analysis is performed with the SAS/STAT procedure TRANSREG (transformation regression). Metric conjoint analysis models are fit using ordinary least squares, and nonmetric conjoint analysis models are fit using an alternating least squares algorithm (Young 1981; Gifi 1990).

Fractional Factorial Design

When the attributes of the product which is to be marketed are finalised and the levels of the attributes to be studied are fixed, the marketer will be having many product combinations. For example, if there are 4 attributes with 3 levels each then 81 product combinations will have to be evaluated for consumer preference which is a tedious task and during a survey the consumers will not be patient enough to spare a long time. Fractional factorial designs can be used to reduce number of product combinations which can be used for the consumer preference study.

Example

This example uses PROC TRANSREG to perform a conjoint analysis to study preferences for fish valued added products. The fish products have four attributes: three with three levels and one with two levels. The attributes are shown in the following table:

Factor	Levels
Main Ingredient	Meat, Fish, Vegetable
Price	Rs 8, Rs 12, Rs15
Product form	Stuffed, Fried, Sandwich
Perceived quality	Yes, No

Generating the Design

We can use the %MktExautocall macro to find a design. When you invoke the %MktEx macro for a simple problem, you only need to specify the numbers of levels and number of runs. The %MktEx macro can create designs in a number of ways. For this problem, it simply looks up an orthogonal design. The following step invokes the %MktEx macro:

```
%MktEx (3 3 3 2, n=18)
```

The first argument to the %MktEx macro is a list of factor levels, and the second is the number of runs (n=18). The %MktEx macro creates two output data sets with the experimental design, Design and Randomized. The Design data set is sorted. In the randomized design, the profiles are presented in a random order and the levels have been randomly reassigned. We use the FORMAT procedure to create descriptive labels for the levels of the attributes. By default, the values of the factors are positive integers.

Data

18 data cards are to be separately prepared and the consumers asked to sort the cards from most preferred to least preferred. The combination numbers (most preferred to least preferred) are entered as data. The data are transposed, going from one observation and 18 variables to 18 observations and one variable named Combo.

The next DATA step creates the variable Rank: 1 for the first and most preferred combination, ..., and 18 for the last and least preferred combination.

The following steps sort the data by combination number and merge them with the design:

```
title 'Product';
data results;
input combo1-combo18;
datalines;
17 6 8 7 10 5 4 16 15 1 11 2 9 14 12 13 3 18;
proc transpose out = results (rename = (col1 = combo)); run;
data results; set results; Rank = _n_; drop _name_; run;
proc sort; by combo; run;
data results(drop=combo);
mergesasuser. dietdes results;
run;
proc print; run;
```

You can use PROC TRANSREG to perform the nonmetric conjoint analysis of the ranks as follows:

```
proctransreg utilities order=formatted separators=',';
model monotone(rank / reflect) =
class (Ingredient PriceFormp_quality / zero=sum);
output out=utils p ireplace;
run;
```

The utilities option displays the part-worth utilities and importance table. The order=formatted option sorts the levels of the attributes by the formatted values. By default, levels are sorted by their internal unformatted values (in this case the integers 1, 2, 3). The model statement names the variable Rank as the dependent variable and specifies a monotone transformation for the nonmetric conjoint analysis. The reflect transformation option is specified with rank data. With rank data, small values mean high preference and large values mean low preference. The reflect transformation option reflects the ranks around their mean (–(rank – mean rank) + mean rank) so that in the results, large part-worth utilities mean high preference. With ranks ranging from 1 to 18, reflect transforms 1 to 18, 2 to 17, ..., r to (19 – r), ..., and 18 to 1. The class specification names the attributes and scales the part-worth utilities to sum to zero within each attribute.

The output statement creates the out= data set, which contains the original variables, transformed variables, and indicator variables. The predicted utilities for all combinations are written to this data set by the p option (for predicted values). The replace option specifies that the transformed independent variables replace the original independent variables, since both are the same.

Consumer Acceptance and Sensory Evaluation

Demand by the time-conscious consumer for convenience foods has increased recently require modern methods in processing, packaging, distributing and advertising. Testing consumer acceptance of a product through standard sensory evaluation methods is adopted by the food industry to determine the general acceptability and shelf life of a product. To predict the acceptability of a product, sensory evaluation is carried out. Taste panel is a group of members specially constituted for sensory evaluation of food products. Desirability and other characteristics of a product will be stated by the taste panel members after tasting the samples of the newly developed food product. Using a prescribed procedure called screening of the taste panel members, the persons who constitute the panel are selected carefully. Descriptive tests, preference tests, difference / discriminatory tests are the statistical methods used to study consumer acceptance.

Descriptive Tests

By using this test, a complete description of all the product characteristics is done. This guides the product developer to modify the product to suit to the needs of the general public. A group of highly trained panelists examine a particular property of a product to provide a detailed descriptive evaluation of it. Appearance has the greatest influence on the consumer since visual appeal stimulates appetite and help to select the product at the first glance among many other brands. Taste of food is also crucial. Once the food has been tasted, the flavour of the food becomes important. It ensures continued use of the product by the consumer. Therefore, it is highly essential that when a new product is developed, consumer acceptance studies of these properties are conducted before promoting the product.

Profile of each characteristic is recorded such as the appearance profile, texture profile, flavour profile, etc. The flavour profile is the description of the taste and odour of the product. The description should contain the intensity of each factor, the order in which the factors are perceived, after taste and overall impression. The texture profile is the description of the textural characteristics perceived in a product, the intensity of each factor and the order in which they are perceived. Mechanical characteristics like hardness, fracturability, chewiness, gumminess, adhesiveness and viscosity and geometrical characteristics like grittiness, coarseness and fibrousness are described.

Scaling: The rating scale method provides the members with a scale showing several degrees of magnitude. The members' task is to assign a rating to a particular attribute such as taste, odour or texture. For each of the property to be evaluated a scale is constructed. For example, if the freshness of the product is to be evaluated, use a 5-point scale such as very fresh, fresh, neutral, spoiled, very spoiled. The panel members together agree upon the scale terms to be used and evaluate the samples. The data are tabulated and analysed using analysis of variance.

Free choice profiling: In this technique, the members are told what attributes are to be judged. Members are presented with samples belonging to the category being examined and are instructed to use a common scale. Each panel member selects the attributes, develops a score card and rate samples accordingly.

Repertory Grid Method: In this method, the panel members are given two or three samples of the same product and asked to describe the similarities and differences in the samples. The information on the attributes thus collected will be analyzed to find out analogies among descriptors obtained.

Demerit system: Various sensory factors associated with different organs of fish like skin, eyes, gills, belly, vent etc. are described and graded from 0 for extremely fresh and above for the extent of spoilage. The number of scores for each factor is given based on its contribution towards spoilage. Every description of the demerit point is very brief usually involving one or two words. No single feature is given undue importance and the fish sample is evaluated based on the totality of the sample.

Preference Tests

Preference tests are affective tests based on a measure of preference from which relative preferences can be determined. Three popular preference tests are paired comparison, the hedonic scale and ranking tests.

Paired comparison test: The member is presented with a pair of samples – one will be the new product developed and another will be a sample of the existing product and asked which he prefers. The member will be instructed what attributes to judge. The two samples of the pair have to be given simultaneously or successively. The time interval between samples of pair may vary from 10 to 40 seconds. Longer intervals are used when stimuli are strong, thus probably highly adapting. When multiple pairs are presented at a session, the interval between pairs should not be less than 40 seconds. The members should rinse their mouth between tasting each sample of a pair. Analysis of data is done by applying t-test or Chi-square test. Using tables for the rapid analysis of paired comparison tests, the data is analyzed (Table 1).

Suppose, the sample was preferred by 16 persons in a panel of 25 members, then we conclude the product is not preferred over the other because at least 18 preferences are needed for a significant result.

Table 1: Number of choices required for significance at various levels in a paired comparison test where either sample may be chosen. Chance probability is 50 per cent and the hypothesis is two-tailed

No. of judg-ments	Minimum Number required			No. of judg-ments	Minimum Num-ber required			No. of judg-ments	Minimum Number required		
	5%	1%	0.1%		5%	1%	0.1%		5%	1%	0.1%
1	....	....	....	31	22	24	25	72	45	48	51
2	....	....	....	32	23	24	26	74	46	49	52
3	....	....	....	33	23	25	27	76	48	50	53
4	....	....	....	34	24	25	27	78	49	51	54
5	....	....	....	35	24	26	28	80	50	52	56
6	6	....		36	25	27	29	82	51	54	57
7	7	....	....	37	25	27	29	84	52	55	58
8	8	8	....	38	26	28	30	86	53	56	59
9	8	9	....	39	27	28	31	88	54	57	60
10	9	10	....	40	27	29	31	90	55	58	61
11	10	11	11	41	28	30	32	92	56	59	63
12	10	11	12	42	28	30	32	94	57	60	64
13	11	12	13	43	29	31	33	96	59	62	65
14	12	13	14	44	29	31	34	98	60	63	66
15	12	13	14	45	30	32	34	100	61	64	67
16	13	14	15	46	31	33	35				
17	13	15	16	47	31	33	36				
18	14	15	17	48	32	34	36				
19	15	16	17	49	32	34	37				
20	15	17	18	50	33	35	37				
21	16	17	19	52	34	36	39				
22	17	18	19	54	35	37	40				
23	17	19	20	56	36	39	41				
24	18	19	21	58	37	40	42				
25	18	20	21	60	39	41	44				
26	19	20	22	62	40	42	45				
27	20	21	23	64	41	43	46				
28	20	22	23	66	42	44	47				
29	21	22	24	68	43	46	48				
30	21	23	25	70	44	47	50				

The nine-point hedonic scale is a popular method used for preference testing. The term hedonic is defined as "having to do with pleasure". Here the member expresses his degree of liking or disliking in the following way:

- Like extremely
- Like very much
- Like moderately
- Like slightly
- Neither like nor dislike
- Dislike slightly
- Dislike moderately
- Dislike very much
- Dislike extremely

The results are analysed using Analysis of Variance. If only two samples are to be compared, the mean scores received by each can be compared using t-test.

Ranking: When ranking for preference, the member is presented with coded samples to rank in order of preference. The results are analysed statistically. Generally, 2-digit or 3-digit codes are given to samples to avoid bias.

Discriminatory or Difference Tests

The important and effective test methods used to measure subjectively the difference between samples are Triangle test and Duo-trio test.

Triangle Test

When a new product developed has to be tested whether it is equal or superior to the existing one, then this test is done. Three coded samples, all coded in 3-digit numbers to eliminate bias of which 2 samples belong to Product A and 1 sample belongs to Product B are provided to the panel members and asked to select the old sample. Now the same test is repeated with 2 samples of Product B and 1 sample of product A. When 10 judgments are there, we get 20 sets of choices for both the products. Using Table 2, the superiority of the Product developed against the existing product can be found out. For example, we use a panel of 25 judges and perform triangular test. We need a score of 23 or 26 to prove the superiority of the new product at 5% or 1%, respectively and the number of judgments here is 50.

Table 2: Number of correct identifications required for significance at various levels in Triangle test. Chance probability is 33.3 per cent and the hypothesis is one-tailed

No. of judgments	Minimum Number required			No. of judgments	Minimum Number required			No. of judgments	Minimum Number required		
	5%	1%	0.1%		5%	1%	0.1%		5%	1%	0.1%
1	....	....	....	31	16	18	20	72	32	34	38
2	....	....	....	32	16	18	20	74	32	35	39
3	3	....	....	33	17	18	21	76	33	36	39
4	4	....	....	34	17	19	21	78	34	37	40
5	4	5	....	35	17	19	22	80	35	38	41
6	5	6	....	36	18	20	22	82	35	38	42
7	5	6	7	37	18	20	22	84	36	39	43
8	6	7	8	38	19	21	23	86	37	40	44
9	6	7	8	39	19	21	23	88	38	41	44
10	7	8	9	40	19	21	24	90	38	42	45
11	7	8	10	41	20	22	24	92	39	42	46
12	8	9	10	42	20	22	25	94	40	43	47
13	8	9	11	43	21	23	25	96	41	44	48
14	9	10	11	44	21	23	26	98	41	45	48
15	9	10	12	45	21	24	26	100	42	46	49
16	9	11	12	46	22	24	27				
17	10	11	13	47	22	24	27				
18	10	12	13	48	22	25	27				
19	11	13	14	49	23	25	28				
20	11	13	14	50	23	26	28				
21	12	13	15	52	24	26	29				
22	12	14	15	54	25	27	30				
23	12	14	16	56	26	28	31				
24	13	15	16	58	26	29	32				
25	13	15	17	60	27	30	33				
26	14	15	17	62	28	30	33				
27	14	16	18	64	29	31	34				
28	15	16	18	66	29	32	35				
29	15	17	19	68	30	33	36				
30	15	17	19	70	31	34	37				

The results of the triangular test indicate whether there is a detectable difference or not between the two samples.

Duo-trio test: Initially, three samples are given to the panel members of which one is the reference sample or control and the other two are test samples. Among the test samples, one is identical with the reference sample. The members are asked to taste and identify the odd sample. In Duo-trio test, the panelist has to judge any difference that can be detected. This test has same applications as that of triangle test, but less efficient because the chance of selecting the correct sample by chance is 50%.

Sensory evaluation is very important in marketing management to predict the acceptability of a product. No laboratory techniques are available to predict the acceptability of a product - it wholly depends on the sensory testing ability of the 'taste panel'. So, care should be taken to constitute taste panels for better predictability.

References

Gifi, 1990. Non-linear multivariate analysis. New York. N.Y., Wiley.

Green, P.E. and Rao, V.R., 1971. Conjoint measurement of quantifying judegmental data. Journal of marketing research 8(3): 355-63.

Luce, R. and Tukey, J.W., 1964. Simultaneous conjoint measurement: A new type of fundamental measurement. Journal of mathematical psychology, 1: 1-27.

Young, F.W., 1990. Quantitative analysis of qualitative data Psychometrika, 46: 357-87.

8

Basics of Time Series Analysis and Forecasting Using Longitudinal Data in Fisheries

Shinoj Parappurathu

Introduction

This chapter is intended to throw light on some statistical and econometric methods used for analysing time series data with specific emphasis on basic concepts related to time series analysis, co-integration techniques as well as forecasting techniques such as ARIMA model. Time series analysis is often helpful in revealing the broad patterns associated with longitudinal data, which in turn would aid in important economic and policy decisions. In fisheries sector too, economic variables such as market demand and supply, prices, etc. are of prime importance and often come in longitudinal format. There are a number of alternative methodologies available in the time series literature to determine relevant parameters pertaining to longitudinal data and the ensuing sections are set apart to delve deeper into this realm.

Some Key Concepts

Stationary Stochastic Process

A random or stochastic process is a collection of random variables ordered in time. A stochastic process is said to be stationary if its mean and variance are constant over time and the value of covariance between two time periods depend only on the distance or the gap between the two time periods and not the actual time at which the covariance is computed. In the time series literature, such a stochastic process is known as a weekly stationary or second order stationarity or covariance stationarity. Let Y_t be a stochastic time series with these properties, then,

Mean: $E(Y_t) = \mu$

Variance: $var(Y_t) = E(Y_t - \mu)^2 = \sigma^2$

Covariance: $\gamma_k = E[(Y_t - \mu)(Y_{t+k} - \mu)]$

White Noise Process

A white noise process is a purely random process which has zero mean, constant variance and is serially uncorrelated, that is u_t is identically distributed as a normal distribution with zero mean and constant variance.

Random Walk Model

It is a purely non-stationary process where, the value of Y at time t is equal to its value at (t-1) plus a random shock.

$$Y_t = Y_{t-1} + u_t$$

A random walk model can be of two types i.e., a random walk without drift and a random walk with drift. There will be a constant term in the random walk with drift.

Integrated Stochastic Process

The random walk model is but a specific case of a more general class of stochastic process known as integrated process. These integrated processes have to be differenced to make them stationary. If the process is integrated of the order one, then first differencing will make it stationary and so on. In general a non-stationary process has to be differenced *d* times if it is integrated of the order *d*.

Tests of Stationarity

There are various tests to find out whether the time series is stationary or not. They are mainly,

1. Graphical Analysis

The time series plotted on a graph by taking time on the x-axis and prices on the y-axis. The graph will give the general trend in the data and from this an idea regarding the mean and variances can be understood. A continuous upward trend in the data suggests that the data is not stationary.

2. Auto Correlation Function and Correlogram

One simple test of stationarity is based on the so-called auto correlation function (ACF). The ACF at lag k is denoted by ρ_k and is defined as

$$\rho_k = \frac{\gamma_k}{\gamma_0} = \frac{\text{Covariance at lag k}}{\text{Variance}}$$

If we plot ρ_k against k the graph we obtain is known as population correlogram. If the correlogram obtained from the data is similar to that of a correlogram of a white noise then the data will be stationary. The correlogram of a non-stationary time-series and the one after differencing is given below.

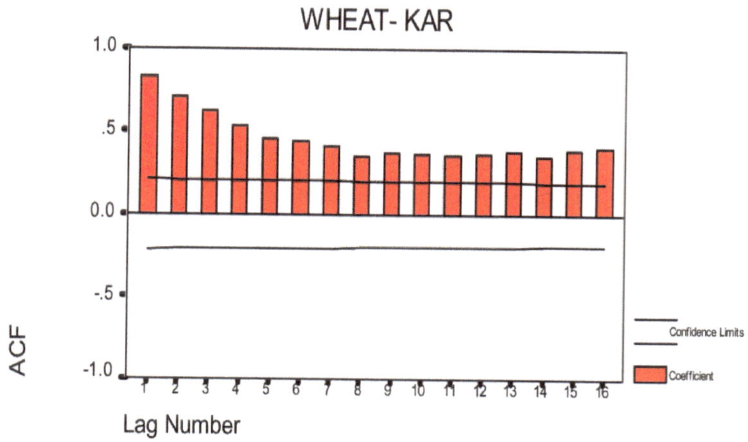

ACF of an undifferenced price series

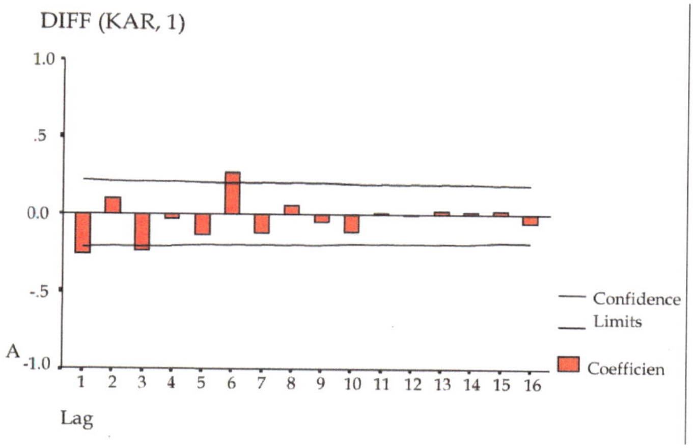

PACF of a differenced price series

3. Unit Root Test

A test of stationarity that has become widely popular over the past several years is unit root test. A random walk model, is given below

$Y_t = \rho Y_{t-1} + u_t$

Here if the value of ρ is equal to one then we can make sure that the series is non-stationary. So in unit root test we regress Y_t on its lagged value and find out if the estimated value of ρ is statistically equal to 1. If it is then Y_t is non-stationary and vice-versa.

Time Series Models for Forecasting

Broadly speaking, there are six approaches to economic forecasting based on time series data: (1) Exponential smoothing methods (2) single-equation regression models (3) simultaneous equation regression model (4) Autoregressive Integrated Moving Average (ARIMA) models and (5) Vector auto regression models. In this paper we mainly discuss ARIMA models.

Co-integration

In economic literature, the implications of drawing economic inferences based on analysis involving non-stationary variables were more or less ignored for a longtime. The irrationality of running spurious regressions with non-stationary time series and drawing important conclusions based on such analysis came to light with the publication of seminal papers by Granger and Newbold (1972) and Nelson and Plosser (1982). Thereafter, diagnostic checks for stationarity have emerged as an important norm before proceeding to any econometric analysis involving time-series variables. In addition, transforming non-stationary variables to stationary ones by successive differencing was mooted as a solution to the problem of non-stationarity. However, several statisticians through their subsequent works questioned the logic of using only the differenced series in economic models as it can potentially cause misspecification errors. Granger (1981) based on empirical examples suggested that a vector of non-stationary variables could have linear combinations which are stationary at levels. Subsequently, Engle and Granger (1987) showed that integrated variables having long-run equilibrium relation with each other can be identified by testing whether the residuals from a regression involving the original variables are stationary or not. This property of time series variables was denoted by the term 'cointegration' and turned out to be a corner stone in subsequent deliberations on the subject. Cointegration analysis, since then is used widely in the economic literature for defining relationships between wide variety of economic variables under varying economic contexts.

Cointegration: Methodological Framework

Let $Y_t = (y_{1t},...,y_{nt})'$ denote an (n×1) vector of I (1) time series. Y_t is cointegrated if there exists an (n×1) vector $\beta = (\beta_1,...,\beta_n)'$ such that,

$$\beta_0 Y_t = \beta_1 Y_{1t} + \ldots + \beta_n Y_{nt} \sim I(0) \quad \ldots (1)$$

In words, the non-stationary time series in Y_t are cointegrated if there is a linear combination of them that is stationary or I (0).

For the sake of simplicity, the above relationship is explained considering the cointegrating relationship of only two variables, i.e., Y_t and X_t hereafter. Two times-series variables are said to be integrated if there exists a long-term equilibrium relationship between them and the degree of their long-run association can be obtained by fitting a classical regression model given by equation (2):

$$Y_t = \beta_0 + \beta_1 X_t + e_t$$

where,

Y_t = Dependent variable

X_t = Independent variable

β_0 = Constant

β_1 = Long-run elasticity, and

e_t =Error-term.

However, assumptions of the classical regression model necessitate that both Y_t and X_t variables should be stationary and the errors should have a zero mean and finite variance. A stationery series is one whose parameters (mean, variance and autocorrelations) are independent of time. As mentioned above, regression between two non-stationary variables may result in spurious relationship with high R^2 and t-statistics that appear to be significant, but with the results of having no economic meaning. Under such circumstances, the series have to be first checked for stationarity. If a time series requires first order differencing to be stationary, then it is said to be I (1) which means integrated of order one. I (2) series requires differencing twice to become stationary and so on. If it is verified that both the series are stationary, then the classical regression model [equation (1)] would hold good and the coefficient would represent the coefficient of price transmission. However, if the two series prove to be non-stationary but integrated of the same order, the validity of regression can be checked by testing the residuals of the regression for stationarity. As demonstrated by Engle and Granger (1987), if the residuals from such a regression turn out to be stationary, then the series are co-integrated and there existed a long-run relationship between the two series. Engle-Granger theorem states that if a set of variables are co-integrated of order (1, 1), then there exists a valid error-correction representation of the data. Converse of this theorem also holds good, that is, if an Error Correction Model (ECM) provides an adequate representation of the variables, then they must be co-integrated. However, if the series are integrated of different orders, the regression equations using such variables would be meaningless and it can be concluded that there cannot exist any long-term relationship between the two.

The stationarity of a series can be tested using a unit root test, the most widely used being Augmented Dickey-Fuller(ADF) unit root test. It would test the null hypothesis that the series has a unit root, i.e. non-stationary.

The test is applied by running the regression of the form given in Equation (3):

$$Y_t = \beta_1 + \delta Y_{t-1} + \alpha_i \sum_{i=1}^{m} \Delta Y_{t-i} + \varepsilon_t$$

where, ε_t is a pure white noise error-term and $\Delta Y_{t-1} = (Y_{t-1} - Y_{t-2})$

Once it is established that the order of integration is the same for the variables of interest, the second stage of testing co-integration can be undertaken. The co-integrating equation is the same as Equation (1). The error-term arising from this regression is then subjected to testing of stationarity. The ADF test in this context is known as Augmented Engle-Granger test whose critical values were provided by Engle and Granger (1987). Davidson and Mackinnon (1993) have revised these values and in the present study, these values have been used. The stationarity in the error-term confirms co-integration between the series and the existence of long-term equilibrium. However, there can be short term disequilibrium which means that a change in series is not immediately passed on to the other. Using the Error Correction Model (ECM), the speed of adjustment towards the long-run path can be ascertained and the model is represented by Equation (4):

$$\Delta Y_t = \alpha_0 + \alpha_1 \Delta X_t + \alpha_2 e_{t-1} + \varepsilon_t$$

where, e_{t-1} is the lagged error-term of the co-integrating regression and ε_t is the disturbance- term. The magnitude of α_2 explains the speed at which the series approaches equilibrium and it is expected to be negative, so that the equilibrium is restored in the long-run.

Econometric Models for Forecasting

Forecasting in agricultural and allied sectors, price forecasting in particular, is a matter of immense importance since it is associated with a variety of uses to farmers, agribusiness firms, and the economy as a whole. Price forecasting helps the farmers to plan for the future farming activities in advance since farm planning and budgeting are considerably dependent upon the future prices. Also it helps the agribusiness firms to foresee the business strategies needed to be taken in the future. Policy decisions regarding exports and imports can also be sorted out easily once the future prices are speculated in advance with a reasonable accuracy. There are various approaches for agricultural price forecasting which are given as follows.

Approach 1: Roulette

In this approach the assumption is that, the future will be the outcome of a pure chance process, and so is inherently unpredictable. This is a very pessimistic type of approach and it underlines that the prices cannot be forecasted accurately and there is no need of price prediction. It is suggested to just guess and hope regarding future prices and hence it is a totally unscientific approach.

Approach 2: Ostrich

It assumes that the future will be identical to the present. So the same trend follows in the future also which is quite easy to extrapolate. Here also there is no scientific back bone to this approach since in the real world price are highly volatile and it is very unlikely that the same trend follows in the future.

Approach 3: Mad Scientist

This approach suggests to find a theoretical cause and effect model i.e., use knowable changes in causes to predict changes in effects. Here all the factors influencing the current prices are identified and they are used as the independent variables to fit a regression model where price is taken as the dependent variable. The coefficients are estimated, and these coefficients are then employed to further find out the future prices.

Approach 4: Trendy

It presupposes that, the past contains patterns (trends) that will continue into the future. Future values are formed by identifying trends and projecting them forwards. So here all the past prices which are having direct or indirect influences on the current prices are identified and a relationship is fitted. The advantage of this approach is that all factors which are likely to affect the future prices will be directly accounted for, through the inclusion of past prices.

There are different econometric models for forecasting the prices based on the above mentioned approaches. The commonly used models are linear models, quadratic models, exponential models, compound growth rate model etc.

1. Linear Regression Models

The general form of the linear regression model is as follows;

$$Y = a + b\,t$$

where, Y= Forecasted prices, t = time period, a and b are the coefficients

2. Non-Linear Regression Models

$$Y = a + b\,t + c\,t^2 + \ldots.. + m\,t^n$$

where, Y= Forecasted prices, t = time period, a, b, c...... m are the coefficients

3. Exponential Models

$$Y = a\,e^{bt}$$

where, Y= Forecasted prices, t = time period, a is the coefficient

4. Time Series Models - ARMA, ARIMA, ARMAX Modeling

In time series models, forecasting of future prices are done using the previous years' prices. The philosophy behind this model is that 'Let the Data speak for themselves '. The general form of time series models is as given below.

$$Y = f(Y^{t-1}, Y^{t-2}, Y^{t-n})$$

Y=Current year prices

$Y^{t-1}, Y^{t-2}, Y^{t-n}$ = Lagged prices

ARIMA (Autoregressive Integrated Moving average) Model

The emphasis of this methodology according to Box and Jenkins, two well-known statisticians, is that on analyzing probabilistic ,or stochastic properties of economic time series on their own under the philosophy 'Let the data speak for them selves'. Unlike regression model which explained the dependent variable is explained by the regressors, this methodology allow Y^t to be explained by past, or lagged values of Y itself and stochastic error terms.

AR, MA, and ARIMA Modeling of Time Series Data

1. An Autoregressive (AR) Process

If Y^t (current price) is dependent upon the previous one year lagged price, then it is called an AR1 Process.

AR(1) Process- $(Y_t - \delta) = \alpha^1 (Y_{t-1} - \delta) + u_t$

Similarly there can be AR2, AR3 and so on upto AR(P) processes.

AR(2) Process-

$$(Y^t - \delta) = \alpha^1 (Y^{t-1} - \delta) + \alpha^2 (Y^{t-2} - \delta) + ut$$

AR(P) Process-

$$(Y_t - \delta) = \alpha_1 (Y_{t-1} - \delta) + \alpha_2 (Y_{t-2} - \delta) + \ldots\ldots. + \alpha_p (Y_{t-p} - \delta) + u_t$$

Where, δ is the mean of Y and u^t is an uncorrelated random error term

2. A moving Average (MA) Process

Here Y at the time t is equal to a constant plus a moving average of the current and past terms. Just as in the case of AR process here also we can have MA1, MA2 up to MA (q) processes.

MA(1) Process - $Y_t = \mu + \beta_0 u_t + \beta_1 u_{t-1}$

MA(2) Process - $Y_t = \mu + \beta_0 u_t + \beta_1 u_{t-1} + \beta_2 u_{t-2}$

MA(q) Process - $Y_t = \mu + \beta_0 u_t + \beta_1 u_{t-1} + \ldots\ldots.. + \beta_q u_{t-q}$

3. An Autoregressive Moving Average (ARMA) Process

When we combine both the AR and MA processes we get the Autoregressive Moving average (ARMA) Process. Here the assumption is that the time series is stationary.

ARMA (1, 1) Process

$Y_t = \theta + \alpha_1 Y_{t-1} + \beta_0 u_t + \beta_1 u_{t-1}$

ARMA (p, q) Process –

$Y_t = \theta + \alpha_1 Y_{t-1} + \ldots\ldots + \alpha_p Y_{t-p} + \beta_0 u_t + \beta_1 u_{t-1} + \ldots\ldots. + \beta_q u_{t-q}$

Where, θ represents a constant term.

4. An Autoregressive Integrated Moving Average (ARIMA) Process

The time series we have already discussed are on the assumption that the time series involved are stationary. Briefly, here the mean and the variance for such a series are constant and the covariance is time invariant. But most of the time series in the real world are non-stationary i.e., they are integrated. Therefore we have to difference the series d times to make it stationary and then apply ARMA (p, q) to it. Then such a process is called an Autoregressive Integrated Moving Average (ARIMA) process of the order p, d and q. It is denoted as ARIMA (p, d, q). If p turns out to be zero the model will reduce to a purely moving average process and if q turns out to be zero the model will reduce to a purely autoregressive model.

5. A Seasonal Autoregressive Integrated Moving Average (SARIMA) Process

Here the assumption is that the data is seasonally non-stationary and has to be differenced according to the order of seasonality to make it stationary. In SARIMA seasonal autoregressive (SAR) and moving average (SMA) factors are also present along with ordinary factors. Such a model is denoted as

SARIMA (p, d, q) (P, D, Q) s Where, P = order of SAR, Q = order of SMA, D = order of seasonal differencing and s = order of seasonality.

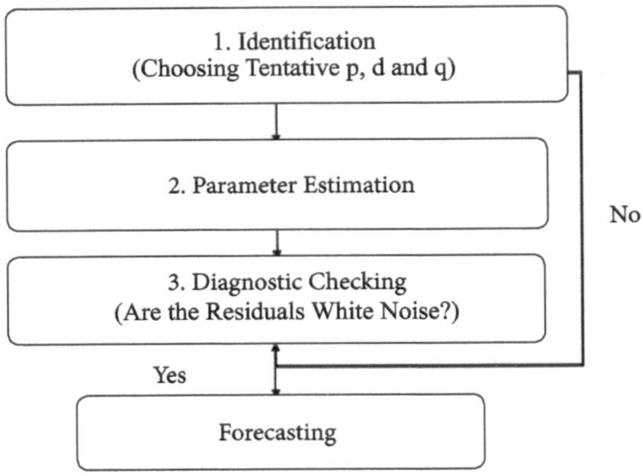

The Box-Jenkins Methodology for ARIMA Model

The B-J methodology in handy in answering the question what are the orders of p, q and d. As shown in the picture above, there are four key steps in this methodology. They are

1. Identification

This is the process to find out the appropriate values of p, d, and q. Two tools i.e., auto correlation function and partial correlation function and the resulting correlograms are commonly used for identification of these orders. The order of p is identified from the partial auto correlation function. Here the number of spikes which crosses the confidence interval level is taken as the tentative order of p. Similarly the number of spikes crossing the confidence interval level in the autocorrelation function is taken as the order of q. The order of d is also identified for the correlograms by checking it for the characters of a white noise series after each differencing operation.

2. Estimation

Having identified the appropriate values of p, d and q the next stage is to estimate the parameters of the autoregressive and moving average terms included in the model. Some times this calculation can be done by simple least squares but some other times we will have to resort to non- linear (parameter) estimation models.

Number of Parameters to be Estimated

Before the estimation can begin, we need to decide on (identify) the specific number and type of ARIMA parameters to be estimated. The major tools used in the identification phase are plots of the series, correlograms of auto correlation (ACF), and partial autocorrelation (PACF). The decision is not straightforward and in less typical cases requires not only experience but also a good deal of experimentation with alternative models (as well as the technical parameters of ARIMA). However, a majority of empirical time series patterns can be sufficiently approximated using one of the 5 basic models that can be identified based on the shape of the auto correlogram (ACF) and partial auto correlogram (PACF). The following brief summary is based on practical recommendations of Pankratz (1983); for additional practical advice, see also Hoff (1983), McCleary and Hay (1980), McDowall, McCleary, Meidinger, and Hay (1980), and Vandaele (1983). Also, note that since the number of parameters (to be estimated) of each kind is almost never greater than 2, it is often practical to try alternative models on the same data.

1. *One autoregressive (p) parameter*: ACF - exponential decay; PACF - spike at lag 1, no correlation for other lags.
2. *Two autoregressive (p) parameters*: ACF - a sine-wave shape pattern or a set of exponential decays; PACF - spikes at lags 1 and 2, no correlation for other lags.
3. *One moving average (q) parameter*: ACF - spike at lag 1, no correlation for other lags; PACF - damps out exponentially.
4. *Two moving average (q) parameters*: ACF - spikes at lags 1 and 2, no correlation for other lags; PACF - a sine-wave shape pattern or a set of exponential decays.
5. *One autoregressive (p) and one moving average (q) parameter*: ACF - exponential decay starting at lag 1; PACF - exponential decay starting at lag 1.

3. Diagnostic Checking

Having chosen the particular ARIMA model, and having estimated its parameters, it has to be seen that whether the model fits the data reasonably well for it is possible that another ARIMA model will the job as well. This is why Box-Jenkins ARIMA modeling is more an art than science; considerable skill is required to choose the right ARIMA model. One simple test of the chosen model is to see if the residuals estimated from the model are white noise; if they are then we can accept the particular fit; if not we must start over. Thus Box- Jenkins methodology is an iterative process. Other two criteria for diagnostic checking are the Akaike Information Criterion (AIC) and Schwarz Information Criterion (SIC).

4. Forecasting

One of the reasons for the popularity of ARIMA modeling is its success in forecasting. In many cases the forecasts obtained by this method are more reliable than those obtained from traditional econometric modeling, particularly for short term forecasts. Forecasting done by extrapolating the model for future prices using the estimated coefficients. Before forecasting, the differenced series has to be integrated to the same order so that the forecasted values are in absolute values. The forecasted value is given by the formula given below.

$$\hat{Y}_{t+1} = \theta + (1+\alpha_1)Y_t + \alpha_1 Y_{t-1} + \ldots\ldots + \alpha_p Y_{p-1} + \beta_1 u_{t-1} + \ldots\ldots \beta_q u_t$$

where $\hat{Y}_{t+1}$ is the predicted price corresponding to the period t+1

5. Evaluation of the Model

Parameter Estimates

You will report approximate *t* values, computed from the parameter standard errors. If not significant, the respective parameter can in most cases be dropped from the model without affecting substantially the overall fit of the model.

Other Quality Criteria

Another straightforward and common measure of the reliability of the model is the accuracy of its forecasts generated based on partial data so that the forecasts can be compared with known (original) observations.

However, a good model should not only provide sufficiently accurate forecasts, it should also be parsimonious and produce statistically independent residuals that contain only noise and no systematic components (e.g., the correlogram of residuals should not reveal any serial dependencies). A good test of the model is (a) to plot the residuals and inspect them for any systematic trends, and (b) to examine the autocorrelogram of residuals (there should be no serial dependency between residuals).

Analysis of Residuals

The major concern here is that the residuals are systematically distributed across the series (e.g., they could be negative in the first part of the series and approach zero in the second part) or that they contain some serial dependency which may suggest that the ARIMA model is inadequate. The analysis of ARIMA residuals constitutes an important test of the model. The estimation procedure assumes that the residual are not (auto-) correlated and that they are normally distributed.

Limitations

The ARIMA method is appropriate only for a time series that is stationary (i.e., its mean, variance, and autocorrelation should be approximately constant through time) and it is recommended that there are at least 50 observations in the input data. It is also assumed that the values of the estimated parameters are constant throughout the series.

Auto-Regressive Model with Exogenous Variable

Multiple variables Auto-Regressive model includes the lagged values of the dependent variable along with one or more exogenous variables as determinants.

$$Y_t = \beta_0 + \beta_1 Y_{t-1} + \beta_2 X_t + e_t$$

Above equation describes a two variables auto-regressive model; these variables are Y and X respectively. Y is an endogenous and X is exogenous variable in the model.

References

Davidson, R. and Mackinnon, J.G., 1993. Estimation and Inference in Econometrics, Oxford University Press, New York, U.S.A.

Engle, R.F. and Granger, C.W.J., 1987. Cointegration and error correction: Representation, estimation and testing, Econometrica, 50: 987-1007.

Granger and Newbold, 1974. Spurious regression in econometrics, Journal of Econometrics, 2: 111-120.

Granger, C.W.J., 1981. Some properties of time series data and their use in econometric model specification, Journal of Econometrics, 23: 123-130.

Gujarati, D.N., 2003. Basic Econometrics, New York: McGraw Hill Book Co. https://www.forecastingsolutions.com/arima.html

Nelson, C.R. and Plosser, C.I., 1982. Trends and random walks in macro-econometric time series: Some evidence and implications, Journal of Monetary Economics, 10, 139-162.

Shinoj P., Kumar, Ganesh B., Sathiadhas R., Datta K.K., Menon, Muktha and Singh, Shiv Kumar, 2008. Spatial price integration and price transmission among major fish markets in India. Agricultural Economics Research Review, 21 (conference number): 327-335.

9

'Resource Cost Ratio' Approach to Estimate Social Profitability of Fisheries Enterprises: The Case of Integrated Fish Farming

Biswajit Debnath

Introduction

Resources must be managed effectively to achieve maximum sustainable agricultural production from limited natural resources. Many of the resources that agriculture depends on (such as water and land) are finite and need to be well managed for long term sustainability. In most cases, this means avoiding adverse effects on the society and environment. With the development of agricultural sectors, the utilization patterns of natural resources in agriculture are important concern for policy makers in present climate change scenario. Unlike other production activities, agricultural activities also utilize factor resources to produce several marketable products or services. The way the natural resources are utilized in agriculture and its concern for society and environment, plays a vital role in economic viability and ecological sustainability of agriculture.

There are several conventional methodologies to estimate the level of profit and efficiency of production system at enterprise level. Only marketable factors are considered in conventional methods keeping the non-marketable component of production system away. Non-marketable component of a production system includes various social and environmental dimensions of it. Nammalwar (1997) has explained the concept of environmentally sound and sustainable enterprises, which include the physical resources, ecological resources, human use value, quality of life values with either way impact of agriculture and environment. In recent years, non-marketable dimensions of agricultural activities have been given importance because of its concern for the environment and society. This chapter provides a methodology that incorporates non-marketable component of agricultural enterprise besides its marketable component. The Resource Cost Ratio (RCR) approach is a modified methodology developed by Debnath *et al.* (2010) from Domestic Resource Cost Ratio (DRCR) that Morris (1990) has applied at macro-economic level to determine the comparative advantage between countries.

Genesis of Resource Cost Ratio (RCR) Approach from Domestic RCR (DRCR)

The RCR approach was developed by Debnath *et al.* (2010) from Domestic Resource Cost Ratio (DRCR) - a simple and comprehensive methodology that is used for macro level policy assessment. DRCR emphasizes on domestic issues of the targeted production sector to assess the trade environmental linkage at global context. This methodology is well appreciated by the economists as it emphasizes on non-tradable issues especially social and environmental factors while reforming or monitoring policy decision.

The versatility of DRCR approach allows it to be used at micro-economic level to examine the economic strength. DRCR methodology is modified as Resource Cost Ratio (RCR) so that the resource cost approach can be used at micro production level in agriculture enterprise. 'Resource Cost Ratio' (RCR) analysis is actually a variant of Domestic Resource Cost Ratio (DRCR) methodology. The DRCR is used to ascertain and measure export / import competitiveness of a particular product vis-à-vis its international counterpart, but RCR reflects the operational efficiency of a production unit. Here, RCR is the ratio of net cost of non-marketable resources and the net value addition effected through marketable route.

In the RCR approach, all possible inputs and outputs are categorized as marketable and non-marketable. Factors of production (inputs) that enter the production system from the market are considered as marketable inputs. These include skilled labor, manure, fertilizers, feed, lime etc. Those which enter the market from the production system are the marketable outputs. These include crops, meat, milk and fish produced on the farm including producer's family consumption. Unskilled labor is a type of non-marketable input for the Resource Cost Ratio approach because there are few employment opportunities for unskilled labor at local level. Biogas slurry or animal waste in integrated farming is considered a non-marketable output because it is reused as manure for agriculture or aquaculture and not sold. Non-marketable outputs also include the environmental and social outputs from agricultural activities.

Valuation of these inputs and outputs must be done at both the producer's price and the economic price. Here, producer's price is the price of input and output items expressed in monetary terms that are paid for or received by actual producers in the local market. The economic price is the real price of input and output items expressed in monetary terms without the effect of all possible externalities.

All inputs and outputs of both marketable and non-marketable categories are structured in a matrix that considered both producer's price and economic (or opportunity) price. The Policy Analysis Matrix (PAM) is used to calculate several indices of RCR (Table 1).Private cost is the sum of the values of both marketable and non-marketable inputs valued at the producer's price, while social cost is calculated by valuing the inputs at their economic values. Private profit was calculated by subtracting the private costs from the sum of the value of marketable and non-marketable outputs, valued at the producer price. Social cost is calculated by adding the economic values of the marketable and non-marketable inputs (G+H). Social profit is calculated by subtracting the social cost from the sum of the marketable and non-marketable outputs (E+F) valued at their economic price. The Resource Cost Ratio (RCR) is calculated by dividing the difference between the economic values of the non-marketable outputs and inputs by the difference between the economic values of the marketable outputs and inputs (E-G). RCR is the ratio of the net cost of non-marketable resources (H-F) and the net value addition from market values (E-G).

Policy Analysis Matrix (PAM) for RCR Approach

In the RCR analysis, the PAM is structured where inputs as well as outputs are classified as marketable and non-marketable. RCR is the ratio of the net cost of non-marketable resources (H-F) and the net value addition from market values.

Table 1: PAM for RCR approach

Item	Outputs		Inputs	
	Marketable	**Non-marketable**	**Marketable**	**Non-marketable**
Producer's price	A (₹)	B (₹)	C (₹)	D (₹)
Economic price	E (₹)	F (₹)	G (₹)	H (₹)
The following indices can be defined from the PAM:				
1. Private profit			= (A + B) – (C + D)	
2. Private profit per unit farm area (ha)			= (1) / farm area (hectare)	
3. Private cost			= (C + D)	
4. Private profit per unit private cost			= (1) / (3)	
5. Social profit			= (E + F) – (G + H)	
6. Social profit per unit farm area (ha)			= (5) / farm area (hectare)	

7.	Social Cost	= (G + H)
8.	Social profit per unit social cost	= (5) / (7)
9.	Resource Cost Ratio (RCR)	= (H – F) / (E - G)

Some of the key inferences of this approach can be explained as follows:

1. Usually, 0 < RCR < 1 on the assumption that (E - G) > 0.
2. Social profit is inversely related to RCR

   Social profit = [(E + F) – (G + H)]

   = [(E – G) – (H – F)]

   = (E – G) [1 – (H – F) / (E - G)]

   = (E – G) [1 – RCR]
3. When Social profit = 0 and (E - G) > 0, then (1 - RCR) = 0 and hence, RCR = 1
4. If RCR = 0, then social profit = (E - G), then assuming (E - G) > 0, social profit > 0 even when RCR = 0. But RCR = 0 implies (H = F) = 0. Thus, when RCR is zero, there is no social profit arising out of value addition through conversion of non-marketable resources into marketable outputs, but social profit may arise because of value addition on marketable inputs i.e., (E - G) > 0.
5. Even when (E - G) < 0, RCR can still be positive if (H - F) < 0. Social profit can be positive even if (E - G) is negative. In this case, negative values from marketable products are more than compensated by positive values of (H - F), non-marketable resources. Investment in rural infrastructure can be of this kind (Datta, 2001).
6. Another interesting situation is where (E - G) > 0, but (H - F) < 0. In this case, value addition through both marketable and non-marketable resources is positive. In this case, social profit is clearly positive, even though the RCR < 0. In other words, a negative value of RCR does not automatically rule out positive social profit.

The Case of Integrated Farming System Spproach in RCR

Integrated farming system refers to a combination of practices that incorporate recycling resources from one farm enterprise to another to optimize production efficiencies. Integrated farming can make a significant contribution to food security for lower income people in the developing world (Mathias, 1994).El Tito (1992) in his comparative study of integrated and conventional crop and vegetable faming systems, went onto define integrated farming systems as ‘an holistic pattern of land use, which integrates natural regulation process into farming activities to achieve a maximum replacement of off-farm inputs and to sustain farm income'.

Integrated farming system based enterprises have many advantages like risk reduction, re-utilization of farm waste, cost reduction and importantly, a perfect farming system approach for adaptation in climate change scenario. The estimation of RCR and its possible sensitivity analysis shows the path for improving social profitability and would explore the possibility of positive environmental benefit through agriculture. Vertical integration approach provides the input materials for next following farming unit, e.g., integration of seed production unit with cultivation system unit. Horizontal integration approach helps in availing some input material for another adjoining production system, e.g., animal waste may be used as manure for fish farming. So, utilization of output or by-product of one production component for another integrated farming component is the key feature of the integrated farming system. Further, product diversification helps in risk reduction in farming as well as to maximize profit either by economies of scale or scope.

In the case of economies of scale and scope, the value of output products tends to increase either because of a large scale of operation or greater product diversification. Possibilities include: (1) decline in the difference between the economic value of non-marketable outputs and inputs, when non-marketable input costs decrease, decline in non-marketable input cost or the value of non-marketable outputs increases; (2) rise in the economic value of the difference between marketable outputs and inputs that occurs with a rise in the marketable output or a fall in the cost of a marketable input. The RCR may fall for integrated farming system because of a combination of these reasons. Addition of a new product like biogas or biogas slurry may result in a less-than proportionate increase in input costs due to the large scale of integration. Further, there may be huge amount of intangible gains, which, though difficult to price and therefore not easy to accommodate in the PAM (without expansion), are extremely important not to lose sight of in an RCR analysis. Any environmental or social benefit from the agriculture production unit, say the environmental benefit from the sewage fed aquaculture system should not be neglected.

Conclusion

Resource cost ratio approach in estimating social profit is versatile in nature. Versatile use of RCR approach can be made through sensitivity analysis. Sensitivity analysis involves partial changes in the values of some important parameters like subsidies (availed for the production), lease rent, intangible outputs (environment related) etc. For this, RCR values for hypothetical cases can be found out and the reasons for robustness of RCR value can be derived. In other words, the approach provides the flexibility to include diverse form of non-marketable factors and innovations in its estimation. Effective management of resources is necessary to achieve maximum sustainable production from limited resources. Besides, Resource management at micro-economic level must be supported by effective approach in farm management and decision making. Resource Cost Ratio approach provides the basis for it. The approach also exposes the scope for incorporating eco-friendly measures in farming system to improve social and environmental benefit especially in the present context of climate change scenario.

Disclaimer: The views or ideas given in this chapter are entirely the view of the authors based on different research activities and the Institute and ICAR are not responsible for any liability arising out of this publication.

References

Borregaard N. and Bradle T., 1999. Towards understanding costs and benefits of trade liberalization – A developing country perspective. In: Report of Workshop of methodologies for environmental assessment of trade liberalization agreements. The Organisation for Economic Co-operation and Development (OECD), Paris.

Datta, S. K., 2001. How to Judge Global competitiveness of Indian Agri-Business - Methodological Issues & lessons for India. In: Implications of WTO Agreements for Indian Agriculture, (Eds. Datta, S. K. and Deodhar, S.). 2001. Oxford & IBH Company, New Delhi. Chapter 19: 571-587.

Debnath, B.,Biradar, R.S., Datta, S.K., Pandey, S.K. and Samal, S.S., 2010. Estimation of private and social profitability of aqua-enterprise using novel resource cost ratio (RCR) approach, Journal of Fisheries Economics and Development, 11 (2): 1-15.

El tito, A. 1992. Integrated farming: An ecological farming approach in European agriculture, Outlook in Agriculture, 21 (1): 33-39.

Mathias, J., 1994. Integrated fish farming in the context of world food security.In: Integrated fish farming, proceedings of the workshop on integrated fish farming. (Eds. Mathias, J., Charles, A.T., and Baotong, H.): 3-17.

Morris, M.L., 1990. Determining Comparative Advantage through DRC Analysis: Guidelines Emerging from CIMMYT's Experience. CIMMYT Economics Paper No. 1. Mexico, D.F.: CIMMYT.

Nammalwar, P., 1997. Environment parameters analysis of aquaculture enterprises –ecological resources. In: Proceedings of the workshop on environmental impact assessment of aquaculture enterprises. Rajiv Gandhi Center for Aquaculture, Chennai: 122-133.

10

Methodologies for Analyzing Fish Marketing Data

Biswajit Debnath

Introduction

The term 'Fish Marketing' is composed of two broad meaning term 'Fish' and 'Marketing'. Before discussing the methodological issues of fish marketing data analysis, we must understand the broader meaning of it as both words are being used frequently with a précised meaning. Google dictionary explains fish as 'A limbless cold-blooded vertebrate animal with gills and fins living wholly in water'. But 'fish' as a component of fish marketing or social science study, means a product and a variety of product, which has different time, space and form utilities for consumers.

The word 'marketing' is not merely buying and selling of fish, rather it connotes a series of activities involved in moving goods or services from point of production to the point of consumption. So 'fish marketing' is defined as physical and institutional set up to perform all activities involved in the flow of fish, fish products and services from the point of initial fish production until they are in the hands of ultimate customers.

It includes assembling, handling, storage, transport, processing, wholesaling, retailing and export of fish and fishery products as well as accompanying supporting services such as market information, establishment of grades and standards, community trade, financing and price risk management and the institutions involved in performing the above functions.

It is desired to have an efficient and well planned distribution networkso that fish and fishery products move from the place of landing or production to the place of consumption at cheapest cost with quality –intact, ensuring the maximum price realization to the farmers or fish producers. It is really challenging to the public agencies involved to meet the requirements of all three major players, viz., fishermen or fish farmers, consumers and traders.

It includes all activities involved in creation of time, place, form and possession utility. Philip Kotler explained marketing as the science and art of exploring, creating, and delivering value to satisfy the needs of a target market at a profit. 'Fish marketing' is different from marketing of other goods and services and even differs a bit from marketing of variousagricultural goods and services. It is due to the specific characteristics of fish as a product of marketing, which are as follows:

1. Fresh fish is highly perishable item and hence its time utility is precious.
2. There are variety of fish and fishery products available which are different in its form and hence, it creates a quite different form utility to the consumers.
3. Lot of variations is observed in the quality of fish and fishery products.
4. Supply of fish and fishery product is irregular.
5. Production is scattered across different regions.
6. Different consumption habits in our country also affect fish marketing activities.
7. International trade in fish marketing is completely different than that of existing marketing within country.
8. A fish or fish product may become an inferior product to even a luxury product based upon the situation exists, which is influenced by consumer's income.

Fish and fishery products are increasingly finding place in the diets of Indian population as a result of their rising income, growing urbanization, unfolding globalization, changing consumption behavior and life style, increasing number of working women population etc. Fish is a healthy food with a few taboos associated with it, unlike meat and the health is associated with high protein, vitamin, minerals and low fat contents.

So 'fish marketing' describes a wide range of concept which is versatile in nature and dealing with methodologies for analyzing fish marketing data also shows a wide variation based upon the type of research objectives and data.

Type of Fish Marketing Research

Study related to fish marketing research could be of three types: 'Exploratory', 'Descriptive' and 'Explanatory' research.

Exploratory research is defined as a research used to investigate a problem which is not clearly defined. It is conducted to have a better understanding of the existing problem, but will not provide conclusive results. For such research, a researcher starts with a general idea and uses this research as a medium to identify issues that can be the focus for future research. An important aspect here is that the researcher should be willing to change his/her direction subject to the revelation of new data or insight. Such a research is usually carried out when the problem is at a preliminary stage. Surveys or polls, interviews, focus groups, observation, online research, literature survey or conducting case study are different approaches followed for exploratory research in fish marketing.

Descriptive research in fish marketing is the next level of research that provides an in-depth description of the phenomenon or population under study. Descriptive research is neither in the category of qualitative research nor in the class of quantitative research, but it uses the features of both types of research. Descriptive research is a suitable choice, if we want to learn about the trends of a particular field or the frequency of an event. This research is also an appropriate option when we do not have any information about the research problem, and primary information gathering is required to establish a hypothesis. Case study, observation and survey are three basic ways to conduct descriptive research in fish marketing.

Explanatory research is the highest form of marketing research that oriented to establish the causes and originate a certain phenomenon. It is a type of quantitative research that discovers the why and the why of a phenomenon. The causes and effects of the studied are revealed from an explanation of the phenomenon deductively from theories or laws. Explanatory research generates operational definitions referring to the phenomenon studied and provides a model closer to the reality of the object of study.

Exploratory and descriptive research in fish marketing deals with the understanding about existing phenomenon as mentioned hereunder in this article. Identifying the market intermediaries, market channels etc. are kind of exploratory and descriptive research in fish marketing.

Exploratory and Descriptive Research Methodologies

Fish Market Structure Conduct Performance (SCP)

There are three important dimensions of marketing system viz., market structure, conduct and performance. Market structure determines the market conduct and performance; the structural characteristics govern the behavior of marketing firms. The market structure has never remained static but kept on changing with the changing environment. An important characteristic of fish marketing in India has been that the private trade dominates the market. With large quantities required to be handled by the private trade, the size and lack of proper markets are getting increasingly crucial.

There are structural weakness of markets like unorganized suppliers as against organized buyers, weak holding capacity of the producers and the perishable nature of the produce in absence of any storage infrastructure. In presence of these characteristics in the market, the rural producers cannot simply be left to fend for themselves so far as marketing of their produce is concerned. In such circumstances, producers will only be losers in the process of development and globalization. Following detailing have been collected from Mebrate and Worku, 2019.

Market Structure

Structure of the Market: The term market structure refers to the number of buyers and sellers, their size distribution, the degree of product differentiation, and the ease of entry of new firms into an industry (Abbott and Makeham, 1981; Bain, 1968; Kohls and Uhl, 1985). Examples of such dimensions include: Degree of Buyers and Sellers Concentration: Number and size distribution of buyers and sellers in the market. Barriers to Potential Entrants: Refers to the relative ease or difficulty with which new dealers may enter into market. Technological, economic, regulatory, institutional, and other factors that inhibit firms from engaging in new businesses or entering new markets, and Degree of Product Differentiation: Refers to the extent to which competing products in a market are differentiated and it is expected to influence the competitive interrelationships of sellers in the market. Market concentration: It can be defined as the number and size of sellers and buyers in the market. Concentration is believed to play a large part in the determination of market behavior within an industry because it affects the interdependence of action among firms. The relationships between concentration and market behavior and performance must not be interpreted in isolation. Other factors, such as firms' objectives, barrier to entry, economies of scale, and assumptions about rival firms' behavior, will be relevant in determining the degree of concentration and relationship between concentration and behavior and performance (Schere, 1980).

Market structure can also be defined as characteristics of the organization of a market, which seem to strategically influence the nature of competition and pricing behavior within the market (Bain, 1968). Structural characteristics may be used as a basis for classifying markets. Markets may be perfectly competitive; monopolistic; or oligopolistic (Scott, 1995). The organizational features of a market should be evaluated in terms of the degree of seller concentration, entry barriers (licensing procedure, lack of capital, and policy barriers), degree of transparency and degree of product differentiation that condition or influence the conduct and strategies of competitors (Wolday, 1994).

Conduct of the Market

Market conduct refers to the behavior of firms or the strategies used by the firms, for example, in their pricing, buying, selling, etc., these activities may require the firms to engage into informal cooperation or collusion (Teka, 2009). Definition of market conduct implies analysis of human behavioral patterns that are not readily identifiable, obtainable, or quantifiable. Thus, in the absence of a theoretical framework for market analysis, there is a tendency to treat conduct variables in a descriptive manner (Thompson, 2011).

Performance of the Market

It is the reflection of the impact of structure and conduct on product price, costs and the volume and quality of output (Cramers and Jensen, 1982). If the market structure in an industry resembles monopoly rather than pure competition, then one expects poor market performance. According to Abbott and Makeham (1981), market performance is how successfully the firm's aims are accomplished, which shows the assessment of how well the process of marketing is carried out.

Marketing Costs

These are incurred to perform various marketing activities in the transportation of goods from producers to consumers. Marketing costs include handling cost (packing and unpacking), costs of searching for a partner with whom to exchange, screening potential trading partners to ascertain their trustworthiness, bargaining with potential trading partners (officials) to reach an agreement, transferring the product, monitoring the agreement to see that its conditions are fulfilled, and enforcing the exchange agreement (Holloway and Ehui, 2002).

Marketing Margin

It is a commonly used measure of the performance of a marketing system (Abbott and Makeham, 1981). It is defined as the difference between the price the consumer pays and the price that is obtained by producers, or as the price of a collection of marketing services, which is the outcome of the demand for and supply of such services (Cramer's and Jensen, 1982). A big margin may result in little or no profit or even a loss for the seller involved depending upon the marketing costs as well as on the selling and buying prices (Mendoza, 1995). Hence, marketing margin is the difference between buying and selling prices. In competitive markets, the margin achieved should be commensurate with the cost of services rendered.

Marketing Channels and Intermediaries (NCAP, 2008)

Fish and fishery products, both fresh and frozen, move in the marketing chain through different channels, domestic or export as the case may be. The marketing channels are distinguished from each other on the basis of market functionaries involved in carrying the produce from the farmers to the ultimate consumers. The length of marketing channel depends on the size of the market, nature of the commodity and pattern of the demand at consumer's level. Fish marketing system in India involve several marketing channels, each with a number of intermediaries between producer and consumer of fish. The common domestic fish marketing channel for marine and freshwater fish in India is one that has auctioneer, commission agent, wholesaler, and retailer as intermediaries.

For Marine and Brackish Water Fish

The fisherman sells his catch through the auctioneer, usually at landing centre, without any value addition. Generally each fisherman has a pre-fixed auctioneer through whom he sells his catch. The auctioneer conducts the auction and sells the fish to local dealers or fish collector, who then takes it to the wholesale market and auction it off to the wholesaler. The auctioneer and fish collector carry out some value addition in terms of cleaning, icing of fish etc. The commission agents operates at the market place and the fish is supplied to the agents who facilitate the disposal of fish to wholesaler and retailers for a fixed commission. The commission agent rarely carries out any value addition. The wholesaler in turn sells his fish to the retailer after re-icing, salting, cleaning, size grading etc. Vendors and hoteliers also might buy their supply of fish from the wholesaler directly. Retailers carry out further value addition by re-icing, size grading, dressing, and filleting the fish and sell the same to the consumers.

For Freshwater Fish

The common fish marketing channel for freshwater fish involves the traders and fishermen cooperative societies. The traders collect the fish from lending centers (either through auction or at fixed prices) and supply the fish to the market for a commission. The fishermen cooperatives carry out the same function as trader. Both the traders as well as the cooperatives sell their fish to the wholesaler. The wholesaler sells the fish to retailer or to mobile trader or bulk consumers like hoteliers. The retailers sell the fish to the consumers after some value addition.

For Export Fish

The common marketing channel for fish export involves the trader or agents and fish processing units. Fish is purchased from the producers by the traders or agents of export processing units. The traders sell the fish to the fish processing unit. After processing, the fish are sent to the respective destinations by the exporters.

Explanatory Research in Fish Marketing

Explanatory Research Methodologies

After the brief discussion made above about the exploratory and descriptive research issues in fish marketing system existing in the country, now we will discuss about the explanatory research methodologies in fish marketing. Before getting into the methodological aspects of data analysis, it is required to understand the process of fish marketing research, especially for explanatory kind of research. Few explanatory type of research in fish marketing may be estimating the demand of fish in a regional area, studying the consumer's perception, estimating income elasticity of demand etc.

Defining research problem, objective of the study and research statement is the primary step in explanatory studies in fish marketing data. After doing so, we would go for review of literature, then formulation of hypothesis, designing research, collection of data, data analysis and finally interpretation of the results and report writing. Methodological issue in data analysis is a part of the whole process. Use of appropriate statistical tool will depend on the preceded steps in the explanatory research. So, we need to be careful, what is the objective of the study and what our research statement is. After collecting data on the marketing aspects, we can choose the suitable tests with a critical observation on the following aspects of the data and variables:

1. Type of data and variables
2. The distribution it follows
3. The research objective studying either a comparison or correlation or causation
4. How many datasets and groups needed in the study design
5. Whether the study design is paired or unpaired type.

After observing above aspects in the data we could get a guidance to go for selecting an appropriate statistical tool for analyzing the data. If our data follows a normal distribution and also fit all the assumption of parametric tests then, we would prefer to go for different parametric tests, whereas, if out data doesn't fulfill the assumption of parametric tests, then we can move for non-parametric tests. For the data that follows dichotomous distribution, we must chose a suitable statistical tools like Mc Nemar tests or Chi-square tests or Logistic or Probit regression. We can follow the guideline mentioned below for choosing between parametric and non-parametric tests (Table 1) as follows:

Observations are independent (Random independent samples), Normally distributed, No outliers, Homogeneity of variances, Sample sizes larger than many of non-parametric tests are the assumption of Parametric tests. Certain assumptions are associated with most non parametric statistical tests, but these are fewer and weaker than those of parametric tests. Parametric tests are more powerful: more likely to detect that truly exists and less likely than parametric tests to make type II error.

Table 1 : Choosing parametric and non-parametric tests for fish marketing data analysis

Particulars	**Parametric**	**Non-parametric**
Assumed distribution	Normal	
Assumed variances	Homogenous	
Typical data	Ratio or Interval	Ordinal or nominal
Dataset relationship	Independent	any
Usual central measure	Mean	Median
Benefits	Can draw more conclusion	Simple, less affected by outliers
	Tests	
Choosing	Choosing parametric tests	Choosing non-parametric tests
Correlation tests	Pearson	Spearman
Independent measures, 2 groups	Independent measure t test	Mann-Whitney test
Independent measures, > 2 groups	One-way, Independent measure ANOVA	Kruskal-Wallis test
Repeated measures, 2 conditions	Matched pair t test	Wilcoxon test
Repeated measures, > 2 conditions	One-way, repeated measure ANOVA	Friedmen's test

For causal studies in fish marketing data analysis, especially for consumer behaviour studies, the methodologies like Probit regression, Conjoint analysis, Rank based Quotient are very useful methods and widely used. Fish demand study, multiple budgeting framework received attention by many researchers in fish demand analysis.

Disclaimer: The views or ideas given in this chapter are entirely the view of the authors based on different research activities and the Institute and ICAR are not responsible for any liability arising out of this publication.

References

Abbot, J.C. and Makeham, J. P., 1981. Agricultural Economics and Marketing in the Tropics, Wing Tai Cheung Printing Co. Ltd, Rome.

Bain, K. and Howells, P., 1988. Understanding Markets: An Introduction and Practice of Marketing, Harvester Wheat Sheaf, London.

Cramer, G.L. and Jensen, W., 1982. Agricultural Economics and Agribusiness, 2nd Edition. McGraw Hill Book Company, USA.

Holloway, G. and Ehui, S., 2002. Expanding Market Participation among Smallholder Livestock Producers: A Collection of Studies Employing Gibbs Sampling and Data from the Ethiopian Highlands. Socio-economic and Policy Research Working Paper 48. ILRI, Nairobi, Kenya.

Kohls, R.L. and Uhl, J. N., 1985. Marketing of Agricultural Product. Fifth Edition. McMillan Publishing Company, NewYork, USA.

Mebrate Y., and Worku, A., 2019. Structure, Conduct and Performance of Fish Market in Central Ethiopia. Management Studies and Economic Systems (MSES), 4 (4): 295-303.

Mendoza, G., 1995. A Primer on Marketing Channels and Margins, G. J. Scott (ends). Prices, Products, and People; Analyzing Agricultural Markets in Developing Countries, Lynne Rienner Publishers, Boulder, London

NCAP, 2008. Exploring market opportunities for fisheries sector in India. Report of the research study. National Centre for Agricultural Economics and Policy Research (presently known as ICAR-National Institute of Agricultural Economics and Policy Research, ICAR-NIAP).

Schere, F.M., 1980. Industrial Market Structure and Economic Performance, 2nd Edition. Rand McNally College Publishing Agency, USA.

Scott, G. J., 1995. Prices, Products and People: Analyzing Agricultural Markets in Developing Countries, Lynne Reinner Publishers, Boulder, London.

Teka, S. G., 2009. Analysis of Fruit and Vegetables Market Chains in Alamata Southern Zone of Tigray: The Case of Onion, Tomato, and Papaya, Unpublished MSc. Thesis, Haramaya University, Haramaya, Ethiopia.

Thompson, C.F. (2011). Swaziland Business Year Book. Christina Forsyth Thompson, MbabaneSwaziland.

Wolday A., 1994. Food Grain Marketing Development in Ethiopia after Reform 1990. A Case Study of Alaba Siraro. The PhD Dissertation Presented to Verlag Koster University, Berlin.

11

Analysis of Classic Breeding Experimental Data Using Linear Mixed Models and Estimation of Variance Components

J. Jayasankar

Setting

The basic difference which dovetails most of the animal breeding experiments is the unbalancedness of the data collated with respect to factorial classes thereby making non-orthogonal nature of the effects involved ubiquitous. Henderson, C.R. (1953), through a series of seminal works dealt with these types of data and came out with three distinct methods to estimate the variance components emanating out of such non-orthogonal animal experiment data. Before discussing those methods, let us delve a bit on the challenges in estimating the variance components in animal breeding data. There are two major reasons, first, several methods of estimation are available (most of which are reduced to the analysis of variance method for balanced data), but not one of them has yet been clearly established as superior to the others. Second, all the methods involve relatively cumbersome algebra; discussion of unbalanced data can therefore easily deteriorate into a welter of symbols, a situation which functions as the anti-appetizer to applied researchers.

A safe starting point towards understanding the nuances of dealing with unbalanced data could be the paper by Henderson (1953) which is quite seminal in nature. The methods described have been often referred to as the gold standard in the area of estimating variance components from unbalanced data for many years to come. Of the three, method 1 is simply an analogue of the analysis of variance method used with balanced data and method 2 is used to correct a deficiency of method 1 that arises out of mixed models and the last one is quite different and involves the method of fitting constants so often used in fixed effects models. Let us see these methods in a bit more detail in the ensuing passages.

Data Considerations

Many of the basic statistical designs are focused towards a successful layout which would enable straight forward comparison between treatments will be valid. There are techniques which are meant for effectively dealing with variable experiments, multiple treatment factors and physical constraints on randomization etc. without sacrificing validity of interpretation. The most convenient setup which satisfies this requirement is the convenient property of being balanced.

Balance is one of the several ways of insuring that our design will have orthogonal, mutually independent, effects. That means the effects are independently estimable. When such situation prevails, then there is no need to worry about possible masking or interference from other variables in the model when we want to assess the importance of one particular variable, or even the interactions among variables.

If all the experiments included only orthogonal effects there would be no need for different types of sum of squares as there would be little room for controversy over how one should interpret an analysis of variance. Although clear cut recommendations in this connection would be difficult, software like SAS have options and directives which have to be clearly understood to be followed.

Balanced and Unbalanced Data

Much of the data generated by animal experiments and studies on other biological resources do not have orthogonal effects. Sometimes the researcher does not get to assign levels of the research variables randomly to subjects, for example, the sex of animal subjects, and is forced to take what experimental subjects are obtainable. Often a model has quantitative covariates that were measured at the same time as the response variable. Such covariables would be orthogonal to the research treatments only by sheer accident. Further, in such studies animals have been known to die in the middle of an experiment, leaving the most well planned experiment lop sided. Such imbalance or non-orthogonality faced during the experimentation phase leads to a lot of statistical retrospection. Suitable amends must be made in the post experiment data visualisation, especially analysis, when such situations arise.

Some design layouts themselves restrict how the treatments may be randomly allotted to the experimental units. The best example for this is the split plot design in which the units are first separated into groups then one set of treatments are randomly allotted to one group of units after which another set of treatments is randomly allotted to the group of units. The analysis performed subsequently, ANOVA, must include multiple error terms. Hence to perform this type of analysis in software like SAS, some tweaking needs to be done like taking the interaction of blocking with whole plot treatment as one level of error to test the first set of treatments.

Another set of challenging data is the repeated measures dataset which are quite common in animal research domain. Balancing in such experimentation is very much an investment which would pay off during the interpretation of analyses.

Data Types and Sum of Squares

Let us consider some data guided splitting of variance as suggested by analytic software like SAS. For a one way completely randomised layout, there is no issue of what type of sum of squares to use, since they are all the same. A two way layout is the simplest design for which issues arise. There are four types of datasets in a two way setup as regards the cell frequencies.

1. Equal cell frequencies: n (i.j) constant for all i and j. This is the so called balanced layout.
2. Proportional cell frequencies: n (i,j) /n (i,l) =n (k,j) /n (k,l) for all i, j, k and l. This is called proportionally balanced layout.
3. Disproportionate, non-zero cell frequencies: n (i,j) > 0 for all i and j, but previous condition does not hold good for some i, j, k, l combinations. This is called unbalanced layout.
4. Empty cells: n(i,j)=0 for some i,j combinations. This layout has got missing cells.

Let us delve a bit in detail the varying impacts of the above mentioned dataset structures using the popular SAS analytics statement. Suppose Proc GLM is used to analyse the above mentioned dataset, with the factors being A and B, the model statement would be

MODEL Y = A B A*B / SS1 SS2 SS3 SS4;

Upon running, SAS would generate four sets sum of squares for testing the interaction and main effects. They are

1. ***Fully sequential:*** Analogous to regression model fitting. A statement A B A*B leading to the computation of SS due to first effect, A, followed by that of B after adjusting A. For some datasets, we could get different results for the statement B A A*B. The only model in which the sum of all effect SS adds up to model SS.
2. ***Method of fitting constants:*** SS for a given effect is adjusted against all other effects except those which have the given effect as a part of

interaction. Hence for A, adjustment would be against B but not A*B as it has A's part in it. Similarly for B, it would be adjusted against A not A*B. A*B would be adjusted against both.

3. ***Weighted squares of means:*** SS for the given effect would be adjusted against all other effects including interactions, which the given effect is a part of.
4. ***Goodnight's method of dealing with missing cells:*** Similar to weighted squares method in spirit, but a strategy is made for the missing cells.

The following table shows which types of SS will be equal for the four kinds of datasets.

Data set type				
Effect	**Balanced**	**Proportional**	**Unbalanced**	**Missing Cell**
A	I=II=III=IV	I=II, III=IV	III=IV	
B	I=II=III=IV	I=II, III=IV	I=II, III=IV	I=II
A*B	I=II=III=IV	I=II=III=IV	I=II=III=IV	I=II=III=IV

As a matter for helping decision making, it can be mentioned that when the data is balanced all the four types of SS give the same result. But for unbalanced data or proportionally balanced data, it is usually advisable to go for Type-III SS. But under special situations, if the situation warrants that interactions need not be adjusted while testing main effects, then Type II SS is better. For split plot type datasets, it is better to use sequential SS (Type I) with due consideration for the sequence of effects. One final thing to remember is choice of SS type will not affect the error SS and hence contrasts won't be affected while testing fixed effects. When analysis is carried out for multiple error terms, then those special error terms have been corrected for the treatment effects they are to be used to test.

Types of Linear Models

The quality of any statistical analysis is best judged by the model that describes the data and it should be conceptualized on three levels viz., true in describing data perfectly, ideal in being close to the true model and operational in being simple and parsimonious. A good operational model is derived from the ideal model. The data quantity and the value of the data points determine the type of operational model and hence set the tone for the subsequent inference.

In the traditional "frequentist" approach fixed and random factors need to be distinguished, whereas in Bayesian approach, there is no such distinction. Fixed factors are those in which the classes comprise all of the possible classes of interest that could be observed. For example, the sex of an animal could be either male, female, sterilized male or sterilized female. Hence, if the number of classes in a factor is small and confined to this number even if conceptual resampling was to be performed an infinite number of times, then the factor is likely to be fixed. Random factors are those whose levels are considered to be drawn randomly from an infinitely large population of levels. For example in a pig experiment, pigs were considered random because the pig population of the world is large enough to be considered infinitely large and the group that was involved in that experiment were a random sample from that population.

Another way to determine if a factor is fixed or random is to know how the results will be used. In a diet experiment, the inferences made is applicable to the diet compositions used in the experiment only and not for others and hence, the factor indicating them is a fixed factor. In contrast, the effect of animals undergoing that trial which are not supposed to have anything peculiar and hence, the animal effects would be a random factor. The differentiation checklist can be summarized as follows:

Particulars	**Random**	**Fixed**
Levels	Selected at random from conceptually infinite collection of possibilities	Finite number of possibilities
Another experiment	Would use different levels from same population	Would use same levels of the factor
Goal	Estimate variance components	Estimate means
Inference	For all levels of the factor (i.e., for population from which levels are selected)	Only for levels actually used in the experiment

Based on the inclusion of fixed and random effects in a liner model, the type of the model is christened viz., fixed, random and mixed, wherein both the types of effects are present and is the most commonly confronted in animal science research.

Mixed Linear Models

The linear model in general comprises three parts:

1. The equation
2. Expectations and variance-covariance structures
3. Assumptions, restrictions and limitations

The Equation

Though the general purpose of this write up is to throw light on the computational possibilities and inferential tips to animal researchers and not to deal with the statistical rigour behind them, for a better definition of framework of discussion some basic statistical discourse is laid out.

The equation of the model defines the factors that may have an effect on the observe trait. A matrix formulation of a general mixed model equation is as follows:

y = Xb + Zu + e

where,

y is the vector of observed values of the trait,

b is a vector of factors, collectively known as fixed effects,

u is a vector of factors known as random effects,

e is a vector of residual terms, which are also random in nature.

X, Z are known matrices, commonly known as design matrices, that describe the precise relationship between the elements of **b** and **u** with those of **y**. When Zu does not appear in the model, it is known as fixed effects model and random model is one where **Xb=1**μ (which is the overall mean effect).

Expectations and VCV matrices

In general terms, the expectations are:

$$E\begin{pmatrix} y \\ u \\ e \end{pmatrix} = \begin{pmatrix} Xb \\ 0 \\ 0 \end{pmatrix}$$

and the variance covariance matrices are

$$V\begin{pmatrix} u \\ e \end{pmatrix} = \begin{pmatrix} G & 0 \\ 0 & R \end{pmatrix}$$

Where, G and R are the general square matrices assumed to have determinant value greater than zero and the elements of which are assumed to be known. Also

V (y) = ZGZ' + R

Assumptions and Limitations

The third part of a model includes items that are not apparent in parts 1 and 2: for example, information about the data or the manner in which the data were collected. Were the animals randomly selected? Did the data arise from many environments, at random? In this part of the model, all those issues must be addressed to. The differences between the ideal and operational model should be listed and possible effects on the analysis should be explained. Hence, this part contains the most important information for assessing the quality of analysis.

Let these be explained with an example formulation:

Suppose there is a progeny data of three sires on temperament scores (on a scale of 1 to 40) taken at milking time as shown below:

Group (CG)	Age	Sire	Score
1	1	1	17
1	2	2	29
1	1	2	34
1	2	3	20
2	2	3	20
2	1	3	20
2	2	1	13
2	1	1	18
2	2	2	25
2	1	2	31

The equation of the model might be

$y_{ijkl} = c_i + a_j + s_k + e_{ijkl}$

where, y_{ijkl} is a temperament score, c_i is a contemporary group effect (CG) which identifies animals that are typically reared and treated alike together; a_j is an age group effect; s_k is the sire effect and e_{ijkl} is the residual effect. Contemporary groups and age groups are often taken to be fixed factors and sires are generally random factors. Age group 1 was for daughters between 18 and 24 months and age group 2 was for daughters between 25 and 32 months of age.

The expectations and variance- covariances are:

$E(y_{ijkl}) = c_i + a_j$

$E(s_k) = 0$

$E(e_{ijkl})=0$

$Var(s_k)=\sigma^2_s$

$Cov(s_k, s_{k'}) = 0$

$Var(e_{ijkl}) = \sigma^2_{ei}$

Thus the residual variance differs between contemporary groups. The sire variance represents one quarter of the additive genetic variance because all progeny are assumed to be half sibs. The sires are assumed to be unrelated.

The assumptions and limitations of the model could be

1. Daughters were approximately in the same stage of lactation when temperament scores were taken,
2. The same person assigned temperament scores for all daughters.
3. The age groupings were appropriate.
4. Sires did not have common ancestry.
5. Sires were mated randomly with dams.
6. Only one offspring per dam.
7. Only one score per daughter.
8. No preferential treatment towards particular daughters.

Matrix formulation

$$y=\begin{pmatrix}17\\29\\34\\16\\20\\24\\13\\18\\25\\31\end{pmatrix},X=\begin{pmatrix}1&0&1&0\\1&0&0&1\\1&0&1&0\\1&0&0&1\\0&1&0&1\\0&1&1&0\\0&1&0&1\\0&1&1&0\\0&1&0&1\\0&1&1&0\end{pmatrix}\text{ and }Z=\begin{pmatrix}1&0&0\\0&1&0\\0&1&0\\0&0&1\\0&0&1\\0&0&1\\1&0&0\\1&0&0\\0&1&0\\0&1&0\end{pmatrix}$$

$$G=\sigma^2_s I_{3x3}$$

$$R=diag\left(\sigma^2_{e1}\sigma^2_{e1}\sigma^2_{e1}\sigma^2_{e1}\sigma^2_{e2}\sigma^2_{e2}\sigma^2_{e2}\sigma^2_{e2}\sigma^2_{e2}\sigma^2_{e2}\right)$$

Henderson's Approach to Analyse Linear Mixed Models

Dr. C.R. Henderson in 1949 as a result of his Ph.D. thesis work at IOWA State University derived a method to approach the linear mixed models (LME) and propounded what is known as method 1. This method was not published until 1953 along with methods 2 and 3 in Biometrics in one of the seminal papers of Henderson.

Method 1

Method 1 superimposes the ANOVA procedure for balanced data to unbalanced data. Method 1 is applicable only when the models are completely random. It is theoretically derived to be unbiased and translation invariant but only under a model of completely random mating and no selection and hence, it is easiest to calculate from quadratic forms (matrix versions of sum of squares).

Method 2

Method 2 requires making amends for the shortcomings of method 1 viz., non-inclusion of genuine fixed factors. The method as advocated by Henderson is unbiased, translation invariant and can be used with mixed models provided that interactions between fixed and random factors or nesting of random factors within fixed factors do not exist in the model. This method requires the least squares (LS) of equations of all factors in the model. A solution vectoris obtained by taking a generalized inverse of the coefficient matrix of the LS equations. It involves the following steps:

1. Construct LS equations for all factors (fixed and random).
2. Adjust **y** for **b** (fixed effect) from the OLS equations of previous step and get the absorbed **y**.
3. Apply method 1 to the adjusted **y** and adjust the coefficients in the expectations of quadratic forms to allow for the changes from the varcov matrix of **y** to adjusted **y**.

Method 3

The starting point of method 3 is the bottleneck experienced in method 2 i.e., the models are not allowed to contain interactions between fixed and random factors or nesting of random factors within fixed factors. This method is the most inclusive of the three and is more computationally intensive and is popularly known as "method of fitting constants" method because reductions in sums of squares due to fitting sub models of the full models are used. This method is unbiased and translation invariant, but not uniquely defined in the sense that more reductions can be computed than are necessary to estimate the variance components. This method has been shown in some cases to provide more accurate estimates of variance components than the other two. The reason behind the accuracy is due to fitting sub models such that two or more random factors are simultaneously considered rather than singly as would be the case in the other two methods. To illustrate this method in simple terms let's assume a model with three random factors A,B and C along with a set of fixed factors denoted by **b**, then following reductions may have to be used.

y'y

R(**b**,A,B,C) (wherein R() notation stands for the reduction in the total sum of squares (TSS) with the fitting of each effect and the letters in parentheses indicate the factors fitted in the respective models, viz. all fixed effects in **b**)

R (**b**, A,B)

R (**b**, A,C)

R (**b**, B,C)

Using the expected values of such reduced sum of squares the estimates of the coefficients and thus thevariance components are arrived at.

Computational Avenues

As indicated in the previous sessions, the methods of analyzing and inferring from these type of data with inherent non-orthogonality is quite computer intensive to say the least. A couple of decades back, there were quite a few options which could attempt to provide solutions to these problems. Ofcourse, generic statistical software like SAS and SPSS were in vogue and had initiated steps to include random effects into their GLM setup, but with lots of restrictions and limitations.

Harvey's LSMLMW & MIXMDL Package

The classic animal breeding experiment demanded the models to have exclusive estimation protocols for the random effects and their variance component after duly compensating for the fixed effects. Hence, a couple of efficient algorithms were developed to attempt to solve the LME using Henderson's MME and method of fitting constants. One such predominant package is the LSMLMW (Least Squares Maximum Likelihood Mixed Weighted) with MIXMDL package for PC version 2 by Walter R. Harvey (1990) which suited the animal breeders the best. The programme written in Fortran language gave options for eleven types of independent variables in LME. They were:

1. Mani and cross-classified effects (A)
2. Nested effects (B:A)
3. Two-way interaction of main effects (A*B)
4. Two-way interaction of main and nested effects (A*C:B)
5. Two-way interaction of nested and main effects (B:A*C)
6. Two-way interaction of nested effects with nested effects (B:A*C:A)
7. Continuous independent variable (X, for covariance analysis)
8. Two-way interaction of main effects with continuous independent variables (A*X)
9. Two-way interaction of continuous independent variables with main effects (X*A)
10. Two-way interaction of nested effects with continuous independent variables (B:A*X)

11. Two-way interaction of continuous independent variables with nested effects (X*B:A)

Further, there are options for fitting polynomials up to 5th degree also available in the package.

General Outline of Model Types

Harvey's LSMLMW and MIXMDL PC-2 has nine model settings in all and the following is a brief description of them.

Model 1 deals with a fully fixed model setup, with only b vector and no u vector.

Model 2 is for the LMEs with one set of cross-classified non-interacting random effects

Model 3 is with one set of non-interacting random nested effects and this is simply the split-plot model

Model 4 deals with one set of cross classified non-interacting random effects and one set of nested non-interacting random effects.

Model 5 models the setup wherein two sets of nested non-interacting random effects exist. The second set of nested random effects is nested within the first set of nested random effects and these are nested within a fixed set of effects.

Model 6 contains one set of cross-classified random effects which interact with one set of fixed cross-classified set of effects. This is the randomized block split-plot design.

Model 7 contains one set of nested random effects which interact with one set of fixed cross-classified effects. This can be alternatively viewed as the nested split-plot design.

Model 8 has one set of cross-classified non-interacting random effects and practically same as model 2, however in this case the MME of Henderson are used to obtain Best Linear Unbiased Predictions (BLUP) of the random effects and Best Linear Unbiased Estimates (BLUE) of the fixed effects and Minimum Norm Quadratic Unbiased Estimates (MINQUE) of the variance components.

and

Model 9 contains one set of non-interacting random nested effects. This is generically similar to model 3 but with the additional options of estimating BLUP of random effects and BLUE for fixed effects and MINQUE for variance components can be estimated.

PROC MIXED of SAS

Coping up with the modern developments in computing generic statistical packages like SAS started developing exclusive routines for analyzing and inferring animal research data. Although the much established PROC GLM was suiting to such data to certain level, the later addition, PROC MIXED exclusively addresses the computational requirements of mixed model data. It is a single tool for analyzing a large array of models used in statistics and it mainly uses the more stable and more efficient procedure of estimation called Residual (Restricted) Maximum Likelihood method or REML in short. Apart from adjusting and absorbing fixed effects and providing avenues for the nesting and cross-classification of the random effects PROC MIXED also provided efficient predictors of functions of random effects viz., BLUP. Further, they also have the option for testing the variance components which have the inherent non-admissibility issue of not to be equal to or less than zero. Hence, the classical methods of testing against the population value of zero doesn't hold good here. As the estimates are based on Likelihoods a sort of Likelihood ratio test is what is required to test random effects. PROC MIXED provides a solution for this problem in the form of a 'Wald" test which can be performed by using the COVTEST option. Further, PROC MIXED can be used for repeated measures data also. Thus this procedure has nice properties in terms of getting reasonable estimates of variance components, computing BLUPs and automatically adjusting fixed effects tests for complex variance structures. Interesting kinds of models that would not be possible to fit without computer intensive methods can now be fit. As a corollary, the two efficient procedures viz., PROC MIXED and PROC GLM can be juxtaposed for comparison and the gist is given below:

Criteria	PROC GLM	PROC MIXED
Nutshell	Designed for **fixed effects** model with allowance for certain adjustments in the presence of random terms. For many designs special attention needs to be given to least square means and contrasts since their standard errors are not necessarily correct. This is true, for example, for split-plot designs.	Designed for mixed effectsmodels. Random terms are incorporated into inference from the outset. Contrasts, least square means and estimates of linear combinations are reported with correct standard errors. Attention needs to be paid to the selection of an appropriate inference space.

Danger	If analysis is driven by accounting for degrees of freedoms and tests, p-values, contrasts, least square means etc. are taken for granted. Depending on the design, some results reported by PROC GLM are actually incorrect.	In order to access the powerful inferential features of PROC MIXED, effects are treated as random when they are actually fixed. Prime example: block effects.
Estimation of fixed effects	Based on Ordinary Least Squares	Based on Generalized Least Squares in a Gaussian error model (estimates are Maximum Likelihood Estimates under normality).
Estimation of variance components	Method-of-Moments estimation (ANOVA method) of solving expected mean squares for the variance components	Maximum Likelihood or Restricted Maximum Likelihood estimation.
MODEL statement	All effects are listed, whether fixed or random	Only the fixed effects are listed. Random effects are listed in the RANDOM statement.
LSMEANS statement	Interprets all effects as fixed, even the random effects. While least square means are correct, their standard errors are not necessarily correct. This is true, even if a RANDOM statement is used.	Only effects in the MODEL statement are assumed fixed. The standard error estimates for least square means account for the random effects.
ESTIMATE statement	Same as LSMEANS statements	Same as LSMEANS statements
Sums of Squares	They are everywhere	Since it is not based on expected mean squares, there are no sums of squares in the standard output.
RANDOM statement	Invokes the calculation of expected mean squares for the listed effects and appropriate test using the /TEST option. The randomness of the effect is not incorporated into the tests of main effects, lsmeans, contrasts, etc.	Signals incorporation of the listed effects in all aspects of inference. Options allow selection of various correlation models describing the dependencies of multiple random effects.
TEST statement	Important statement to utilize correct error terms in testing model effects. Prime examples: Subsampling designs, split-plot designs	Gone.

References

C.R. Henderson Applications of Linear Models in Animal Breeding University of Guelph, Guelph, on Third Edition edited by L.R. Schaeffer 1984.

Harvey, W.R., 1990. User's guide for LSMLMW and MIXMDL. Mimeo, Ohio State Univ. Columbus. 91pp.

Henderson C.R., 1953. Estimation of variance and covariance components. Biometrics, 9: 226-252.

Little, R.C., Milliken, G.A., Stroup, W.W., Wolfinger, R.D., 1996. SAS System for Mixed Models. Cary, NC: SAS Institute Inc. 633 pp.

Schaeffer Lawrence R., 2010. Linear Models and Animal Breeding, Centre for Genetic Improvement of Livestock, Department of Animal and Poultry Science, University of Guelph, Guelph, Norway

Searle S.R., 1969. Another look at Henderson's methods of estimating variance components. Biometrics 24: 749-788.

Searle S.R., Casella, G., and Mc Culloch, C.E., 1992. Variance Components. New York: Wiley.

12

Linear, Non-Linear and Logistic Regression in Fisheries Data Analysis

Ramasubramanian V.

Introduction

Regression analysis is one of the most widely used techniques for studying relationships involving multiple variables for analysing data by expressing a relationship between a variable of interest (the response) and a set of related predictor variables. The regression models include linear and non-linear approaches assuming appropriate functional forms. A good account on regression analysis and related topics can be found in Draper and Smith (1998), Montgomery *et al.* (2001), Chatterjee and Hadi (2006) etc. In this write-up, regression model fitting, some of the detection techniques which are useful in detecting the problem of multicollinearity between the so-called 'independent variables' and also outlier detection in data are discussed. Linear regression with qualitative regressor variables is also discussed. Moreover, variable selection procedures, goodness of fit measures for model adequacy and validation are also discussed. In addition, non-linear regressions viz., logistic (both binary and multinomial) when the response variable is qualitative are also covered.

Multiple Linear Regression Modeling

Let the response variable (variable to be forecasted) be denoted by Y and the set of predictor variables, by X_1 , X_2 , ..., X_p , where p denotes the number of predictor variables. The true relationship between Y and (X_1, X_2 ,...,X_p) can be approximated by a multiple linear regression model given by

$$Y=\beta_0 +\beta_1 X_1 +\beta_2 X_2 +.... +\beta_p X_p + \varepsilon.$$

Here β_0 and β_i (i=1, 2,...,p) are parameters to be estimated and ε is random error. Some assumptions are made about this model like the relationship of the response Y to the predictors X_1, X_2,..., X_p is linear in the regression parameters β_0, β_1,..., β_p , the errors are assumed to be independently and identically distributed (iid) normal random variables with mean zero and a common variance σ , the errors are independent of each other (their pair-wise covariances are zero) and that the predictor variables $X_1, X_2, \ldots, X_p$ are non-random and measured without error.

Francis Galton (in 1880's) coined the term 'regression' to refer to tall parents tending to beget offsprings which were not taller than their parents (also it was noted that parents who were short in height got children who were somewhat taller than their shorter parents). Thus, it was observed that mean filial regression towards mediocrity was directly proportional to the parental deviation from it. The cases in point were size of seedlings, Parents height (X) & Child's height (Y) etc. In today's model fitting situations, no "regression" in original sense but the same term prevailed. Thus regression analysis meant the average relationship between variables studied.

Adequacy and validation of the regression models need to be checked and for that several measures are available in the literature. Many assumptions has to hold good in regression analysis such as the relationship between y and x's is linear, the errors have zero mean and constant variance, the errors are uncorrelated, the errors are normally distributed etc. For checking whether these assumptions are adequately satisfied by any fitted regression model residual analysis is resorted to. Residuals are nothing but differences between the observations and the corresponding fitted values. Residuals have zero mean and approximate average variance as "Error Mean Sum of Squares" from the regression ANOVA. Sometimes the standardised residuals are used. Residual plots are useful for detecting validity of assumptions on errors and model adequacy. Some important residual plots for detecting model inadequacies are stated in brief. Plot of residuals against fitted values. If this plot indicates that the residuals can be contained in a horizontal band, then there are no obvious model defects. If not so, transformations on the regressor and/or the response variable may be required. Plot of standardised residuals against independent variable with no apparent trend as evidence of correct model specification.

Lack of normality and non-constant error variance frequently go hand in hand. Fortunately, it is often the case that the same transformation that helps stabilize the variance is also helpful in approximately normalizing the error terms. It is therefore, desirable that the transformation for stabilizing the error variance be utilized first, and then the residuals studied to see if serious departures from normality are still present.

Given a set of many regressors (or factors that affect the study variable), a subset of most important variables need to be selected before proceeding for fitting any regression model. Variable selection is the process of determining the appropriate subset of that should be used in the model given a pool of candidate regressors that are the possible influential factors. For variable selection, either resort to subset regression models (all possible regressions) or use one of the three stepwise regression methods. viz., stepwise selection, forward selection, backward selection. The criteria for evaluating subset regression models and hence the adequacy of a regression model are the coefficient of multiple determination R^2, adjusted R^2, Residual Mean Square, Mallow's C_p statistic and the PRediction Error Sum of Squares (PRESS).

Multiple linear regression when some regressors are qualitative is tackled in a different manner. This can be done by using dummy or indicator variables for the qualitative regressor variables. If there are 's' levels of a qualitative regressor variables, then (s-1) dummy variables need to be used. One of the levels need to be mentioned as base or reference category. For e.g., if a qualitative regressor variable has four levels, the following dummy variables D1, D2 and D3 are used taking values for the corresponding levels as follows (if the first level has been taken as base):

	D1	D2	D3
Level 1	0	0	0
Level 2	1	0	0
Level 3	0	1	0
Level 4	0	0	1

Detection of violations of assumptions in regression model, particularly multicollinearity need also to be done with respect to independence of the so called 'independent variables' or regressors. When there are near linear dependencies among regressors, then the problem of multicollinearity is said to exist. The variable pool may not contain the right variables in the proper functional forms and we may have included variables with a high degree of multicollinearity, which may cause problems in estimation, prediction and interpretation. Strong multicollinearity between independent variables results in large variances and co-variances for the least squares estimators of the regression coefficients. Multicollinearity also tends to produce least squares estimates of regression coefficients that are too large in absolute value. Some of the measures for detection of multicollinearity are simple pairwise correlations, 'variance inflation factors' and Eigen system analysis of the correlation matrix of the regressors. In order to pinpoint which variables contribute for the greater effect of multicollinearity, 'Belsley's procedure is also employed. Estimation methods such as ridge regression and principal components regression in place of ordinary least squares regression is specifically resorted to combat the problems induced by multicollinearity. However, these procedures yield biased estimators of regression coefficients. In addition, outlier detection in data used for multiple linear regression can be carried out by using measures such as Hat matrix, WSSD, Cook's distance, DFFITS, DFBETAS, COVRATIO.

Non-Linear Regression

A 'non-linear model' is one in which at least one of the parameters appear nonlinearly. More formally, in a 'non-linear model', at least one derivative with respect to a parameter should involve that parameter. Some examples of a nonlinear model are:

$$Y(t) = \exp(a\,t + b\,t^2)$$

$$Y(t) = at + \exp(-bt)$$

Sometimes, the term 'intrinsically linear' is used to indicate a nonlinear model which can be transformed to a linear model by means of some transformation. For example, the first model given above is 'intrinsically linear' in view of the logarithmic transformation which makes it linear. The second model, however, cannot be converted into a form which is linear in the parameters. Such a model is said to be intrinsically nonlinear.

Non-linear regression models may be more appropriate when definite information is available about the form of the relationship that exists between the response and the predictor variable mostly time. In particular, trends or growth patterns of agricultural parameters over time are sometimes non-linear leading to what are called 'growth models' like logistic, Gompertz, mono-molecular, Richards (which empasses the previous three) etc. Processes producing sigmoidal or elongated S-shaped curves are quite common in agricultural data. A most common method of non-linear estimation is Levenberg-Marquardt's method.

Logistic Regression

In case, our response variable in model is qualitative in nature, then probabilities of falling this response variable in various categories can be modeled in place of response variable itself, using same model but there are number of constraints in terms of assumptions of multiple regression model. First, since the range of probability is between 0 and 1, whereas, right hand side function in case of multiple regression models is unbounded. Second, error term of the model can take only limited values and error variance are not constants but depends on probability of falling response variable in a particular category.

Generally, conventional theory of multiple linear regression (MLR) analysis has been applied for a quantitative response variable, while for the qualitative response variable or more specifically for binary response variable, it is better to consider alternative models like logistic regression. Usually discriminant analysis could be used for addressing each of the above problems. However, because the independent variables are mixture of categorical and continuous variables, the multivariate normality assumption may not hold. Structural relationship among various qualitative variables in the population can be quantified using number of alternative techniques. In these techniques, primary interest lies on dependent factor which is dependent on other independent factors. In these cases, the most preferable technique is either probit or logistic regression analysis as it does not make any assumptions about the distribution of the independent variables. The dependent factor is known as response factor. In this model building process, various log odds related to response factors are modeled. As a special case, if response factor has only two categories with probabilities p_1 and p_2 respectively, then the odds of getting category one is (p_1 / p_2). If $\log (p_1 / p_2)$ is modeled using ANalysis Of VAriance (ANOVA) type of model, it is called logit model. Again, if the same model is being treated as regression type model, then it is called logistic regression model. In a real sense, logit and logistic are names of transformations. In case of logit transformation, a number p between values 0 and 1 is transformed with $\log \{p/(1-p)\}$, whereas, in case of logistic transformation a number x between $-\infty$ to $+\infty$ is transformed with $\{e^x /(1 + e^x)\}$ function. It can be seen that these two transformation are reverse of each other i.e., if logit transformation is applied on logistic transformation function, it provides value x and similarly, if logistic transformation is applied to logit transformation function, it provides value p. Apart from logit or logistic regression models, other techniques such as CART i.e., Classification and Regression Trees can also be used to address such classification problems. A good account of literature on logistic regression are available, to cite one, Klienbaum (1994).

For discussing about the violation of assumptions of linear regression model when response is qualitative, to start with, linear regression is considered in order to explain the constraints in using such model when the response variable is qualitative. Consider the following simple linear regression model with single predictor variable and a binary response variable:

$Y_i = \beta_0 + \beta_1 X_i + \varepsilon_i$, i = 1, 2, …, n

where the outcome Y_i is binary (taking values 0,1), $\varepsilon_i \sim N(0, \sigma_\varepsilon^2)$, and are independent and n is the number of observations.

Let π_i denote the probability that $Y_i = 1$ when $X_i = x$, i.e.

$\pi_i = P(Y_i = 1 | X_i = x) = P(Y_i = 1)$

thus $P(Y_i = 0) = 1 - \pi_i \rightleftharpoons$

Under the assumption $E(\varepsilon_i) = 0$, the expected value of the response variable is

$E(Y_i) = 1.(\pi_i) + 0.(1 - \pi_i) = \pi_i$

If the response is binary, then the error terms can take on two values, namely,

$\varepsilon_i = 1 - \pi_i$ when $Y_i = 1$

$\varepsilon_i = 1 - \pi_i$ when $Y_i = 0$

Because the error is dichotomous (discrete), normality assumption is violated. Moreover, the error variance is given by:

$$V(\varepsilon_i) = \pi_i(1-\pi_i)^2 + (1-\pi_i)(-\pi_i)^2$$
$$= \pi_i(1-\pi_i)$$

It can be seen that variance is a function of π_1's and it is not constant. Therefore the assumption of homoscedasticity (equal variance) does not hold.

Logistic regression is normally recommended when the independent variables do not satisfy the multivariate normality assumption and at the same time the response variable is qualitative. Situations where the response variable is qualitative and independent variables are mixture of categorical and continuous variables, are quite common and occur extensively in statistical applications in agriculture, medical science etc. The statistical model preferred for the analysis of such binary (dichotomous) responses is the binary logistic regression model, developed primarily by a researcher named Cox during the late 1950s. Processes producing sigmoidal or elongated S-shaped curves are quite common in agricultural data. Logistic regression models are more appropriate when response variable is qualitative and a non-linear relationship can be established between the response variable and the qualitative and quantitative factors affecting it. It addresses the same questions that discriminant function analysis and multiple regression do but with no distributional assumptions on the predictors. In logistic regression model, the predictors need not have to be normally distributed, the relationship between response and predictors need not be linear or the observations need not have equal variance in each group etc. A good account on logistic regression can be found in Fox (1984) and Kleinbaum (1994).

The problem of non-normality and heteroscedasticity (see section 2) leads to the non-applicability of least square estimation for the linear probability model. Weighted least square estimation, when used as an alternative, can cause the fitted values not constrained to the interval (0, 1) and therefore cannot be interpreted as probabilities. Moreover, some of the error variance may come out to be negative. One solution to this problem is simply to constrain π to the unit interval while retaining the linear relation between π and regressor X within the interval. Thus,

$$\pi = \begin{cases} 0 & ,\beta_0+\beta_1X<0 \\ \beta_0+\beta_1X & ,0\le\beta_0+\beta_1X\le 1 \\ 1 & ,\beta_0+\beta_1X>1 \end{cases}$$

However, this constrained linear probability model has certain unattractive features such as abrupt changes in slope at the extremes 0 and 1 making it hard for fitting the same on data. A smoother relation between π and X is generally more sensible. To correct this problem, a positive monotone (i.e. non-decreasing) function is required to transform $(\beta_0 + \beta_1 x_i)$ to unit interval. Any cumulative probability distribution function (CDF) P, meets this requirement. That is, respecify the model as $\pi i = P(\beta_0 + \beta_1 x_i)$. Moreover, it is advantageous if P is strictly increasing, for then, the transformation is one-to-one, so that model can be rewritten as $P^{-1}(\pi i) = (\beta 0 + \beta 1 xi)$, where P^{-1} is the inverse of the CDF P. Thus the non-linear model for itself will become both smooth and symmetric, approaching $\pi = 0$ and $\pi = 1$ as asymptotes. Thereafter, maximum likelihood method of estimation can be employed for model fitting.

The Logistic response function resembles a S-shape curve, a sketch of which is given in the following figure. Here the probability π initially increases slowly with increase in X, and then the increase accelerates, finally stabilizes, but does not increase beyond 1.

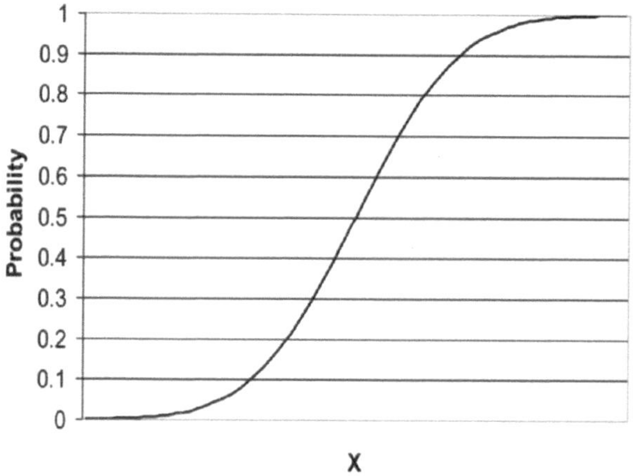

The shape of the S-curve can be reproduced if the probabilities can be modeled with only one predictor variable as follows:

$\pi = P(Y=1|X=x) = 1/(1+e^{-z})$

Where, $z = \beta_0 + \beta_1 x$, and e is the base of the natural logarithm. Thus, for more than one (say r) explanatory variables, the probability π is modeled as

$\pi = P(Y=1|X_1=x_1 \ldots X_r=x_r)$

$\quad = 1/(1+e^{-z})$

where, $z = \beta_0+\beta_1 x_1+\ldots+\beta_r x_r$

This equation is called the logistic regression equation. It is nonlinear in the parameters β_0, $\beta_1 \ldots$ β_r. Modelling the response probabilities by the logistic distribution and estimating the parameters of the model constitutes fitting a logistic regression. The method of estimation generally used is the maximum likelihood estimation method.

To explain the popularity of logistic regression, let us consider the mathematical form on which the logistic model is based. This function, called f (z), is given by

$f(z) = 1/(1+e^{-z}), -\infty < z < \infty$

Now when $z = -\infty$, $f(z) = 0$ and when $z = \infty$, $f(z) = 1$. Thus the range of f (z) is 0 to1. So the logistic model is popular because the logistic function, on which the model is based, provides

- Estimates that lie in the range between zero and one.
- An appealing S-shaped description of the combined effect of several explanatory variables on the probability of an event.

Maximum Likelihood Method of Estimation of Logistic Regression parameters are discussed briefly. For simplicity, a simple binary logistic regression model with only one explanatory variable is considered. The model is given by

$$\pi_i = P(Y_i = 1|X_i = x_i) = 1/(1+e^{-Z})$$

where $z = \beta_0 + \beta_1 x_i$, and e is the base of the natural logarithm. The binary response variable Y_i takes only two values (say 0 and 1). Since each Y_i observation is an ordinary Bernoulli random variable, where:

$$P(Y_i = 1) = \pi_i$$

and $P(Y_i = 1) = 1 - \pi_i$,

the probability distribution function is represented as follows:

$$fi(Yi) = \pi_i^{Yi}(1\pi_i)^{1-Yi}, Yi = 0,1, i = 1,2,\ldots,n$$

Since Y_i's are independent, then the joint probability density function is:

$$g(Y_1 \ldots Y_n) = \prod_{i=1}^{n} f_i(Yi) = \prod_{i=1}^{n} \pi_i^{Yi}(1-\pi_i)^{1-Yi}$$

$$\log_e g(Y_1 \ldots Y_n) = \log_e \prod_{i=1}^{n} \pi_i^{Yi}(1-\pi_i)^{1-Yi}$$

$$= \sum_{i=1}^{n} Y_i \log_e \pi_i / (1-\pi_i)$$

Since E $(Y_i) = \pi_i$, for a binary variable it follows that

$$1-\pi_i = \left[1+e^{-(\beta_0+\beta_1 X_i)}\right]^{-1}$$

Then,

$$\log_e\left[\pi_i / (1-\pi_i)\right] = \beta_0 + \beta_1 X_i$$

Hence the log likelihood function can be expressed as follows:

$$\log_e L(\beta_0,\beta_1) = \sum_{i-1}^{n} Y_i(\beta_0+\beta_1 X_i) - \sum_{i-1}^{n} \log_e\left[1+e^{-(\beta_0+\beta_1 X_i)}\right]$$

Where, L (β_0, β_1) replaces g $(Y_1 \ldots Y_n)$ to show explicitly that the function can now be viewed as the likelihood function of the parameters to be estimated, given the sample observations.

The maximum likelihood estimates β_0 and β_1 in the simple logistic regression model are those values of β_0 and β_1 that maximize the log-likelihood function. No closed-form solution exists for the values of β_0 and β_1 that maximize the log-likelihood function. Computer intensive numerical search procedures are therefore required to find the maximum likelihood estimates $\hat{\beta}_0$ and $\hat{\beta}_1$. Once these estimates $\hat{\beta}_0$ and $\hat{}$ are found, by substituting these values into the response function the fitted response function, say, $\hat{\pi}_i$, can be obtained. The fitted response function is as follows:

$$\hat{\pi}_i = \left(\frac{1}{1+e^{-(\hat{\beta}0+\hat{\beta}1Xi)}}\right)$$

When log of the odds of occurrence of any event is considered using a logistic regression model, it becomes a case of logit analysis. Here the thus formed logit model will have its right hand side as a linear regression equation.

The model validation can be done by employing various tests on any fitted logistic regression model. The tests related to the significance of the estimated parameters, goodness of fit and predictive ability of the models are discussed subsequently.

Wald and Likelihood ratio tests are the commonly used tests for testing the overall significance of the logistic regression model.

Wald test: Let $\hat{\beta}$ be the vector of parameter estimates obtained. Let a set of restrictions be imposed in the form of a hypothesis H_0: $\beta = 0$. If the restrictions are valid, then at least approximately $\hat{\beta}$ should satisfy them. The Wald statistic is then defined as

$$W = \hat{\beta}'\left[\text{Var}(\hat{\beta})\right]^{-1}\hat{\beta}$$

Under H_0, in large samples, W has a Chi-square distribution with degrees of freedom equal to the number of restrictions imposed.

Likelihood Ratio (LR) Test: The LR statistic is defined as two times the logarithm of the ratio of the likelihood functions of two different models evaluated at their MLEs. The LR statistic is used for testing the overall significance of the model. Assuming that there are r_1 variables in the model under consideration which can be considered as the full model, based on the MLEs of the full model, L (full) is calculated. Besides this, the likelihood function L (reduced) is calculated for the constant only model. The LR statistic is then defined as:

$$LR= -2\ [\text{In}\{L\ (\text{reduced})\} - \text{In}\ \{L\ (\text{full})\}]$$

LR is asymptotically distributed as Chi-square with degrees of freedom equal to the difference between the number of parameters estimated in the two models.

Comparison between various logistic regression models fitted and with other classification methods such as discriminant function and decision tree methods can be made with respect to their classifying ability with the help of (2 x 2) classification tables in case of a binary response group variable. The columns are the two observed values of the dependent variable, while the rows are the two predicted values of the dependent. In a perfect model, all cases will be on the diagonal and the overall percent correct will be 100%.

Critical Terms Associated with Classification Table are as Follows

Hit rate: Number of correct predictions divided by sample size. The hit rate for the model should be compared to the hit rate for the classification table for the constant-only model.

Sensitivity: Percent of correct predictions in the reference category (usually 1) of the dependent. It also refers to the ability of the model to classify an event correctly.

Specificity: Percent of correct predictions in the given category (usually 0) of the dependent. It also refers to ability of the model to classify a non event correctly.

False positive rate: It is the proportion of predicted event responses that were observed as nonevents

False negative rate: It is the proportion of predicted nonevent responses that were observed as events.

Higher the sensitivity and specificity lower the false positive rate and false negative rate, better the classificatory ability.

Further concepts under logistic regression that deal with association between attributes/ variables are discussed here. An association exists between two variables if the distribution of one variable changes when the level (or value) of the other variable changes. If there is no association, the distribution of the first variable is the same regardless of the level of other variable. Odds ratio is usually used for measuring such associations. For example, consider the following table having two attributes 'Weather' and 'Mood of boss' each at two levels.

Particulars	**Mood of boss**	
	Good	**Bad**
Weather		
Rain	82	18
Shine	60	40

Odds of an event is the ratio of the probability of an event occurring to the probability of it not occurring. That is,

Odds=P(event)/{1-P(event)} = P(event=1)/P(event=0)

In the above table, there is 82% probability that the mood of the boss will be 'Good' in case of 'Rain'. The odds of 'Good mood' in 'Rain' category =0.82/0.18 =4.5. The odds of 'Good mood' in 'Shine' category =0.60/0.40 =1.5. The odds ratio of 'Rain' to 'Shine' equals (4.5/1.5) =3 indicating that the odds of getting 'Boss in good mood' during 'Rain' is three times those during 'Shine'. Also there is 18% probability that mood of boss will be 'Bad' in case of 'Rain'; the odds of 'Bad mood' in 'Rain' =0.18/0.82 =0.22. Thus, in case the probability is very small (0.18 in this case), there is no appreciable difference in mentioning the same as probability or odds.

The importance of odds ratio is case of logistic regression modeling can be further explained by taking a simple case of influence of an attribute "Gender" X with two levels (Male or Female) on another attribute "opinion towards legalized abortion" Y with two levels (Yes=1, No=0). Logistic regression when written in its linearised form takes the following 'logit' form:

$$logit\ \{Y=1 | X=x\} = \log (\pi/ (1-\pi))=\log (odds) = \beta 0+\beta 1 * x$$

Now,

Odds (Females)= exp(*β0+β1*) and Odds (Males)= exp(*β0*). Hence

Odds ratio = exp(*β0+β1*)/exp(*β0*) = *exp(β1)*.

Thus here regression coefficient of Y on X i.e. *β1* is not directly interpreted but after taking exponentiation of it.

Multinomial logistic regression modeling is also discussed subsequently for the sake of completeness.

Let X is a vector of explanatory variables and π denotes the probability of binary response variable then logistic model is given by

$$\log \text{it}\,(\pi) = \log\left(\frac{\pi}{1-\pi}\right) = a = X\beta = g(\pi)$$

where, 'alpha' is the intercept parameter and 'beta' is a vector of slope parameters. In case response variable has ordinal categories say 1, 2, 3,-------, I, I+1 then generally logistic model is fitted with common slope based on cumulative probabilities of response categories instead of individual probabilities. This provides parallel lines of regression model with following form

$$g[\Pr\text{ob}(y \le i(\times))] = \alpha_i + x\beta, 1 \le i \le I$$

where, α_1, α_2, ..., α_k, are k intercept parameters and β is the vector of slope parameters.

In addition, ordinal logistic regression taking qualitative response variable with three categories (which have an order or gradation), for simplicity is given by

$$\text{logit}\,[\Pr(Y \le j - 1 / X)] = \alpha_j + \beta^T X\,,\ j = 1,2$$

where α_j are two intercept parameters ($\alpha_1 < \beta_2$), $\beta^T = (\beta_1, \beta_2, \ldots\ldots, \beta_k)$ is the slope parameter vector not including the intercept terms, $X^T = (X_1, X_2, \ldots, X_k)$ is vector of explanatory variables. This model fits a common slope cumulative model i.e. 'parallel lines' regression model based on the cumulative probabilities of the response categories. logit (π_1) and logit ($\pi_1 + \pi_2$) are given by

$$\log\left(\frac{\pi_1}{1-\pi_1}\right) = \alpha_1 + \beta_1 X_1 + \beta_2 X_2 + \ldots\ldots + \beta_k X_k\,,$$

$$\log\left(\frac{\pi_1+\pi_2}{1-\pi_1-\pi_2}\right) = \alpha_2 + \beta_1 X_1 + \beta_2 X_2 + \ldots\ldots + \beta_k X_k \text{ respectively.}$$

where,

$$\pi_1(X) = \frac{e^{\alpha 1+\beta^T X}}{1+e^{\alpha 1+\beta^T X}}$$

$$\pi_1(X) + \pi_2(X) = \frac{e^{\alpha 2+\beta^T X}}{1+e^{\alpha 2+\beta^T X}}$$

with $\pi_1 + \pi_2 + \pi_3 = 1$

π_j (X) denotes classification probabilities Pr (Y=j-1 / X) of response variable Y, j = 1,2,3, at X^T.

These models can be fitted through maximum likelihood procedure.

References

Chatterjee, S. and Hadi, A.S. (2006). Regression analysis by example, 4th Edition, John Wiley and Sons, New Jersey.

Draper, N. R. and Smith, H. (1998). Applied Regression Analysis, 3rd edition. New York: Wiley.

Kleinbaum, D.G. (1994). Logistic regression: A self learning text, New York: Springer.

Montgomery, D.C., Peck, E.A. and Vining, G.G. (2001). Introduction to Linear Regression Analysis. 3rd edition, John Wiley and Sons, New Delhi.

Seber, G.A.F. and Wild, C.J. (1989): Nonlinear regression, New York: John Wiley.

13

Market Integration and Price Transmission Analysis of Time Series Data in Fisheries

Krishnan M.

Abstract

The definition of market integration and price transmission, the usefulness and limitations of time series econometrics and the interpretation of the estimated parameters are examined in this article. Completeness, dynamics and asymmetry of adjustment and a testing framework is applied and the extent of price transmission is defined in terms of notional components with each component being tested individually. It is argued that, collectively, these tests can be utilized to assess market integration and the extent to which policies and other distortions affect price transmission. The testing framework is applied to Indian shrimp trade.

Introduction

A fundamental issue when analysing trade policy reform in global seafood market is the extent to which domestic agricultural commodity markets in developing countries respond to changes in international prices. Price transmission from the world to domestic markets is central in understanding the extent of the integration of economic agents into the market process.

Studies on the transmission of price signals are founded on concepts related to competitive pricing behavior. In spatial terms, the classical paradigm of the Law of One Price, as well as the predictions on market integration provided by the standard spatial price determination models (Enke, 1951; Samuelson, 1952; Takayama and Judge, 1972) postulate that price transmission is complete with equilibrium prices of a commodity sold on competitive foreign and domestic markets differing only by transfer costs, when converted to a common currency. These models predict that changes in supply and demand conditions in one market will affect trade and therefore prices in other markets as equilibrium is restored through spatial arbitrage.

The absence of market integration, or of complete pass-through of price changes from one market to another has important implications for economic welfare. Incomplete price transmission arising either due to trade and other policies, or due to transaction costs such as poor transport and communication infrastructure, results in a reduction in the price information available to economic agents and consequently may lead to decisions that contribute to inefficient outcomes. Fisheries and seafood food trade policy reform, especially, is a priority issue in the next WTO negotiations, as trade liberalization is viewed as encouraging allocative efficiency and long run growth.

Price transmission studies are ostensibly an empirical exercise testing the predictions of economic theory and providing important insights as to how changes in one market are transmitted to another, thus reflecting the extent of market integration, as well as the extent to which markets function efficiently. In addition to the body of research and application that tests economic theory, price transmission mechanisms feature prominently in all global fisheries and aquaculture partial equilibrium models, such as the World Food Model of the UN Food and Agriculture Organization and other models such as that of developed by Tyers and Anderson (1992). In these models, the price transmission parameter values consist of key building blocks and play an important role in determining the direction, magnitude and distribution of welfare effects of trade policy scenarios (for a review of price transmission mechanisms in partial equilibrium models see Sharma, 2002). Given the increasing use of these models to address sensitive policy issues, such as trade liberalization and the distribution of benefits and costs across countries and population groups, there is an urgency to review these mechanisms and fine-tune them for further applications.

Literature Review and Main Findings

Several authors have studied price transmission within the context of the Law of One Price (*inter alia* Ardeni, 1989; Baffes, 1991) or within the context of market integration (Ravallion, 1986; Sexton *et al.*, 1991; Palaskas and Harriss 1993; Zanias, 1993; Gardner and Brooks, 1994; Blauch 1997). The concept and the analytical techniques have also been used to evaluate policy reform, such as *ex post* assessment of market integration in the context of the implementation of the structural adjustment programmes (Goletti and Babu, 1994; Alexander and Wyeth, 1994; Dercon, 1995). Another vein of research focuses on vertical price transmission along the supply chain from the consumer to the producer level (see for example Brorsen *et al.*, 1985; Wohlgenant, 1985; Kinnucan and Forker, 1987; Shroeter and Azzam, 1991; Goodwin and Holt, 1999; Prakash 1998; von Cramon-Taubadel, 1999).

The large body of research on market integration and price transmission, both spatially and vertically, has applied different quantitative techniques and has highlighted several factors that impede the pass-through of price signals. Distortions introduced by governments in the form of policies either at the border, or as price support mechanisms weaken the link between the international and domestic markets. Fisheries policy instruments such as import tariffs, tariff rate quotas, and export subsidies or taxes, intervention mechanisms, as well as exchange rate policies insulate the domestic markets and hinder the full transmission of international price signals by affecting the excess demand or supply schedules of domestic commodity markets (Gardner, 1975; Mundlak and Larson, 1992; Quiroz and Soto, 1996; Baffes and Ajwad, 2001; Abdulai, 2000; Sharma, 2002).

In theory, spatial price determination models suggest that, if two markets are linked by trade in a free market regime, excess demand or supply shocks in one market will have an equal impact on price in both markets. The implementation of import tariffs, in general, will allow international price changes to be fully transmitted to domestic markets in relative terms. Thus a proportional increase in the international price will result in an equal proportional increase in the domestic price, at all points in time provided that tariff levels remain unchanged. However, if the tariff level is prohibitively high, changes in the international price would be only partly, if at all, transmitted to the domestic market, as domestic prices may be close to the autarky price level, thus obliterating opportunities for spatial arbitrage and resulting in the two prices moving independently of each other, as if an import ban was implemented. Other policy instruments such as tariff rate quotas may result in international price changes not being at all points of time proportionately transmitted to domestic prices, as changes in the domestic price level will depend on two different tariff rates that are applied according to whether the volume of imports falls within or outside the quota level. In the event that imports are equal to the quota level, changes in the international price may not affect the domestic price level at all, provided that these changes are relatively small, as compared to the difference between the within-the-quota and the out-of-the-quota tariff levels. The implementation of price support policies, such as intervention mechanisms and floor prices, may result in the international and the domestic price being completely unrelated or being related in a non linear manner, depending on the level of the intervention or floor price relative to the international price. Changes in the international price will have no effect on the domestic price level when the international price lies on a level lower than that to which the floor price has been set. However, any changes in the international price above the floor price level will be transmitted to the domestic market.

Thus, floor price policies may result in the domestic price being completely unrelated to the international market below a certain threshold determined by the floor price, or in the two prices being related in a non-linear manner with increase in the international price being fully transmitted to the domestic level, whilst decreases are slowly and incompletely passed-through.

Apart from policies, domestic markets can also be partly insulated by large marketing margins that arise due to high transfer costs. Especially in developing countries, poor infrastructure, transport and communication services give rise to large marketing margins due to high costs of delivering the locally produced commodity to the border for export or the imported commodity to the domestic market for consumption. High transfer costs and marketing margins hinder the transmission of price signals, as they may prohibit arbitrage (Sexton, Kling and Carman, 1991; Badiane and Shively, 1998). As a consequence, changes in world market prices are not fully transmitted to domestic prices, resulting in economic agents adjusting (if at all) partly to shifts in world supply and demand.

Non-competitive behaviour such as that considered in pricing-to-market models (Dornbush, 1987; Froot and Klempeter, 1989; Krugman, 1986) can hinder market integration. Pricing-to-market models postulate that firms may absorb part of exchange rate movements by altering export prices measured in home currency in order to retain their market share. Alternatively, oligopolistic behaviour and collusion among domestic traders may retain price differences between international and domestic prices in levels higher that those determined by transfer costs.

Most of the studies utilize time series econometric analysis techniques that test for the co-movement of prices. The development of these techniques, which include cointegration and error correction models, has become the standard tool for analysing spatial market relationships, replacing earlier empirical tools, such as the bivariate correlation coefficient and regressions. Nevertheless, time series analysis has also being criticized as unreliable (Blauch, 1997; Barrett and Li, 2002) with recent research focusing on switching regime models that incorporate data on prices, volumes traded and transactions costs. The debate on the application methodology for testing for market integration and price transmission has a relatively long history starting with Harriss (1979). Blauch (1997) provides a review of the debate and examines the statistical performance of econometric tests for market integration. In essence, linear tests for market integration and price transmission are thought of as crude and inappropriate (Blauch, 1997; McNew, 1996; McNew and Fackler, 1997;

Fackler and Goodwin, 2002 and Barrett and Li, 2002). Non linearity in market relationships that arise from arbitrage conditions, unsynchronized price cycles, discontinuous trade and non-stationary transfer costs are thought of as rendering linear representations and models not useful and inaccurate.

Although there is some merit in the above criticisms, especially as far as non-stationary transfer costs are concerned, time series analysis can provide useful insights into the issue of market integration and price transmission if an appropriate testing framework is employed and the results are interpreted correctly. Market integration is formally testable, if one adheres to the definition implied by the standard spatial equilibrium model. However, the *extent* of price transmission is an inherently ambiguous concept. Cointegration and error correction models provide an analytical tool that can focus beyond the case of market integration or complete price transmission, in testing notions such as completeness, speed, and asymmetry of the relationship between prices. For example, discontinuities in trade, within a time series modeling framework, correspond to slow speed of convergence to a long run relationship, whilst non-linearity may be modeled as asymmetric responses to price changes. Time series models have small data requirements as compared to other methodologies, relying on price series only, which are more easily available for developing countries. In addition, time series applications perform a useful role in signaling potential failures in markets and in contributing to the assessment of the direction, magnitude and distribution of welfare effects of trade policy reforms. However, it is important to note that, in general, time series applications may also founder while attempting to achieve an unattainable goal, that of giving an universal measure of the extent of price transmission in terms of a single parameter or test.

Price Transmission Notions and Components and the Impact of Food Policies

Given prices for a commodity in two spatially separated markets p_{1t} and p_{2t}, the Law of One Price and the Enke-Samuelson-Takayama-Judge model postulate that at all points of time, allowing for transfer costs *c*, for transporting the commodity from market 1 to market 2, the relationship between the prices is as follows:

$$p_{1t} = p_{2t} + c \qquad \text{... (1)}$$

If a relationship between two prices, such as (1), holds, the markets can be said to be integrated. However, this extreme case may be unlikely to occur, especially in the short run. At the other end of the spectrum, if the joint distribution of two prices were found to be completely independent, then one might feel comfortable saying that there is no market integration and no price transmission. In general, spatial arbitrage is expected to ensure that prices of a commodity will differ by an amount that is at most equal to the transfer costs with the relationship between the prices being identified as the following inequality:

$p_{2t} - p_{1t}$ £ c ... (2)

Fackler and Goodwin (2001) refer to the above relationship as the spatial arbitrage condition and postulate that it identifies a weak form of the Law of One Price, the strong form being characterized by equality (1). They also emphasize that relationship (2) represents an equilibrium condition. Observed prices may diverge from relationship (1), but spatial arbitrage will cause the difference between the two prices to move towards the transfer cost. The spatial arbitrage condition implies that market integration lends itself to a cointegration interpretation with its presence being evaluated by means of cointegration tests. Cointegration can be thought of as the empirical counterpart of the theoretical notion of a long run equilibrium relationship. If two spatially separated price series are cointegrated, there is a tendency for them to co-move in the long run according to a linear relationship. In the short run, the prices may drift apart, as shocks in one market may not be instantaneously transmitted to other markets or due to delays in transport, however, arbitration opportunities ensure that these divergences from the underlying long run (equilibrium) relationship are transitory and not permanent.

The spatial arbitrage condition encompasses price relationships that lie between the two extreme cases of the strong form of the Law of One Price and the absence of market integration. Depending on market characteristics, or the distortions to which markets are subject, the two price series may behave in a plethora of ways, having quite complex relationships with prices adjusting less than completely, or slowly rather than instantaneously and according to various dynamic structures or being related in a non-linear manner. Given the wide range of ways prices may be related, the concept of price transmission can be thought of as being based on three notions, or components (Prakash, 1998; Balcombe and Morisson, 2002). These are:

- Co-movement and completeness of adjustment which implies that changes in prices in one market are fully transmitted to the other at all points of time;
- Dynamics and speed of adjustment which implies the process by, and rate at which, changes in prices in one market are filtered to the other market or levels; and,

- Asymmetry of response which implies that upward and downward movements in the price in one market are symmetrically or asymmetrically transmitted to the other. Both the extent of completeness and the speed of the adjustment can be asymmetric.

Within this context, complete price transmission between two spatially separated markets is defined as a situation where changes in one price are completely and instantaneously transmitted to the other price, as postulated by the Law of One Price presented by relationship (1). In this case, spatially separated markets are integrated. In addition, this definition implies that if price changes are not passed-through instantaneously, but after some time, price transmission is incomplete in the short run, but complete in the long run, as implied by the spatial arbitrage condition. The distinction between short run and long run price transmission is important and the speed by which prices adjust to their long run relationship is essential in understanding the extent to which markets are integrated in the short run. Changes in the price at one market may need some time to be transmitted to other markets for various reasons, such as policies, the number of stages in marketing and the corresponding contractual arrangements between economic agents, storage and inventory holding, delays caused in transportation or processing, or "price-levelling" practices.

Asymmetric response of one price to another implies non-linear adjustment and deserves some further discussion. Many researchers have worked on the issue of asymmetric price responses utilizing the asymmetric error correction model developed by Granger and Lee (1989) or Threshold cointegration models proposed by Enders and Granger (1998). Abdulai (2000) provides a comprehensive discussion on the rationale behind spatial asymmetric price response. In addition to policies, market power is often cited as a source of asymmetries (Scherer and Ross, 1990). Industry concentration and imperfectly competitive behaviour beyond the farm-gate implies that wholesalers, or middlemen with power over price, may exercise pricing strategies that result in a slow and incomplete pass-through of increases in the international price and a fast and complete transmission of decreases in the international price to prices upstream, as their margins are squeezed.

However, in the short run, asymmetric price transmission may also occur for reasons other than policies and market power. In spatial markets, inventory holding behaviour in domestic markets may lead to asymmetries, as high international price expectations lead to stock accumulation. The subsequent release of stocks, *post* the realization of high international price expectations, may exert downward pressure on the domestic market and cause the domestic price not to rise as much as it would in the absence of inventories (Maccini,

1978; Blinder, 1982). Other reasons for asymmetric price adjustment include different reaction to increases and decreases of input costs, depending on whether prices are rising or falling, as competition between wholesalers with high fixed costs and excess capacity may result in producer prices that increase rapidly when demand for processed product is high, but decrease at a slower rate when demand is low (Bailey and Brorsen, 1989; Kovenock and Widows, 1998). Therefore, it is important to note that although most researchers agree that asymmetric price response may be due to concentration and non-competitive pricing behaviour, the theoretical underpinnings of this hypothesis (see for example Wohlgenant, 1999), as well as the related empirical evidence are inconclusive. For example, Griffith and Piggot (1994) detected asymmetries in the Australian lamb and beef markets, but not in the pork market in spite of the fact that the Australian pork market is more concentrated than the other meat markets.

A Price Transmission Testing Framework

As mentioned in section 2, the extent of price transmission lacks a direct unambiguous empirical counterpart in the form of single formal testing. The definition of price transmission provided in the section above encompasses the case of perfect market integration, the inherent dynamic market relationships that arise due to inertia or discontinuities in trade, as well as non-linearity that may arise due to policies and other distortions in arbitrage. More importantly, it implies hypotheses, through its components, that are testable within a cointegration-error correction model framework. A number of time series techniques can be used to test each of the components of price transmission and thus ultimately assess the extent of price transmission. These are as follows:

- Cointegration;
- Causality;
- Error correction mechanism; and,
- Symmetry

Each of the above tests is taken to present evidence about the components of transmission thus providing particular insights into its nature. Collectively, these techniques offer a framework for the assessment of price transmission and market integration.

The concept of cointegration (Granger, 1981) and the methods for estimating a cointegrated relation or system (*inter alia* Engle and Granger, 1987; Johansen, 1988, 1991, 1995) provide a framework for estimating and testing for long run equilibrium relationships between non stationary integrated variables. [49] Cointegration has been extensively discussed and applied in the literature and thus a detailed examination is beyond the scope of this paper (Maddala and Kim, 1998 provide a thorough and extensive review of cointegration). However, a brief description of the concept and the estimation methods in the context of the present analysis is provided.

If two prices in spatially separated markets (or different levels of the supply chain) p_{1t} and p_{2t} contain stochastic trends and are integrated of the same order, say I(*d*), the prices are said to be cointegrated if:

$$p_{1t} - b\,p_{2t} = u_t \qquad \text{... (3)}$$

is I(0).

b is referred to as the cointegrating vector (in the case of two variables a scalar), whilst equation (3) is said to be the cointegrating regression. The above relationship can be estimated utilizing *inter alia* Ordinary Least Squares OLS (Engle and Granger, 1987), or a Full Information Maximum Likelihood method developed by Johansen (1988, 1991) that is most commonly encountered in the literature. More specifically, p_{1t} and p_{2t} are cointegrated, if there is a linear combination between them that does not have a stochastic trend even though the individual series contain stochastic trends (see Stock and Watson, 1988, for the stochastic trend representation of cointegrated systems). Cointegration implies that these prices move closely together in the long run, although in the short run they may drift apart, and thus is consistent with the concept of market integration. Engle and Granger test the null of no cointegration by applying unit root tests on $\hat{u}_t$. Johansen derived the distribution of two test statistics for the null of no cointegration referred to as the Trace and the Eigenvalue tests.

As u_t is stationary, the prices contain stochastic trends that have a long-run proportionality, with the cointegrating parameter *b* measuring the long-run equilibrium relationship between them. This parameter has sometimes been interpreted as the "elasticity of price transmission", when the price series are converted into logarithms. However, this cointegrating parameter does not identify this elasticity, or in other words, the completeness of transmission, particularly well, as recognized by Balcombe and Morrison (2002) and Barrett and Li (2002). Cointegration is a statistical concept and thus "atheoretical", whilst the cointegrating parameter may not have any economic interpretation, in the way a parameter of a structural model has. For example, if prices in spatially separated markets have a common stochastic trend reflecting inflation, the cointegrating parameter will be equal to one mirroring a proportionality of unity and implying that price transmission is complete.

Nevertheless, failure to reject the null of non cointegration implies that the two prices drift apart in the long run, as they are driven by stochastic trends that are not proportional. In this case, some changes in one price, say the international market price, may to a certain extent be transmitted to the domestic market price, however, other factors, such as policies or deviations from marginal cost pricing determine the movements of the domestic market price, thus resulting in absence of market integration. A potential shortcoming of cointegration in testing for market integration is the implicit assumption that transfer costs are stationary (Fackler and Goodwin, 2001; Barret and Li, 2002). Non stationary transfer costs will result in cointegration tests suggesting the absence of market integration, as the international and domestic prices drift apart, in spite of the fact that price signals are transmitted from one market to another. Nevertheless, non-stationary transfer costs cause domestic prices to move independently from international prices, thus limiting the information that is available to producers.

In addition to formally testing market integration, the concept of cointegration has an important implication, purported by the Granger Representation Theorem (Engle and Granger, 1987). According to this theorem, if two trending, say I (1), variables are cointegrated, their relationship may be validly described by an Error Correction Model (ECM), and vice versa (see also brief description in the Annex). In the case that prices from two spatially separated markets, p_{1t} and p_{2t}, are cointegrated, the Vector Error Correction (or VECM) representation is as follows:

$$\begin{pmatrix}\Delta p_{1t}\\ \Delta p_{2t}\end{pmatrix} = \begin{pmatrix}\mu_1\\ \mu_2\end{pmatrix} + \begin{pmatrix}\alpha_1\\ \alpha_2\end{pmatrix}(p_{1t-1}-\beta p_{2t-1}) + A_2 \begin{pmatrix}\Delta p_{1t-1}\\ \Delta p_{2t-1}\end{pmatrix} + \dots + A_k \begin{pmatrix}\Delta p_{1t-k}\\ \Delta p_{2t-k}\end{pmatrix} + \begin{pmatrix}v_{1t}\\ v_{2t}\end{pmatrix} \quad \dots (4)$$

where v_{1t} and v_{2t} are *iid* disturbances with zero mean and constant finite variance, whilst the operator D denotes that the I(1) variables have been differenced in order to achieve stationarity.

The inclusion of the levels of the variables, p_{1t} and p_{2t} alongside their differenced terms Dp_{1t} and Dp_{2t} is central to the concept of the ECM. Parameters contained in matrices $A_2 \dots A_k$, measure the short run effects, while b is the cointegrating parameter that characterizes the long run equilibrium relationship between the two prices. The levels of the variables enter the ECM combined as the single entity $(p_{1t-1} - \beta p_{2t-1})$ which reflects the errors or any divergence from this equilibrium, and correspond to the lagged error term of equation (3).

The vector $\begin{pmatrix}\alpha_1\\ \alpha_2\end{pmatrix}$ contains parameters, usually, $0<|\alpha_i|<1$ $i=1,2$, commonly called error correction coefficients, that measure the extent of corrections of the errors that the market initiates by adjusting p_{1t} and p_{2t} towards restoring the long run equilibrium relationship. The speed with which the market returns to its equilibrium depends on the proximity of a_i to one. Within this context, short run adjustments are directed by, and consistent with, the long run equilibrium relationship, allowing the researcher to assess the speed of adjustment that shapes the relationship between the two prices.

In the context of market integration and price transmission studies, the ECM, as well as its further applications discussed below, is perhaps the most useful tool as it provides a stylized picture of the relationship between two prices. The model provides a structure within which gradual, rather than instantaneous price transmission can be tested, thus taking into account discontinuities in trade and other factors that may impede market integration over time. Most importantly, the proximity of the error correction coefficient to -1 can be used to assess the extent to which policies, transaction costs and other distortions delay full adjustment to the long run equilibrium. Sharma (2002) in a paper aiming to assess market integration between several Asian wheat markets and the world market, estimated ECMs and conducted an extensive policy review. His findings suggest that in countries such as Pakistan, India, Sri Lanka and Indonesia, where government intervenes in the domestic market through various policy instruments, the error correction coefficients were estimated to lie between -0.01 and -0.07 indicating a slow adjustment to the long run relationship.

Another important implication of cointegration and the error correction representation is that cointegration between two variables implies the existence of causality (in the Granger sense) between them in at least one direction (Granger, 1988). The definition of causality and its relevance in the context of market integration and price transmission warrants some discussion. Cointegration itself cannot be used to make inferences about the direction of causation between the variables, and thus causality tests are necessary. Granger (1969) proposed an empirical definition of causality based only on its forecasting content: if x_t causes y_t then y_{t+1} is better forecast if the information in x_t is used, since there will be a smaller variance of forecast error. This definition has caused considerable controversy in the literature (see for example Pagan, 1989) as it really indicates precedence, rather than instantaneous causality that most economists profess. Nevertheless, if two markets are integrated, the price in one market, p_1, would commonly be found to Granger-cause the price in the other market, p_2 and/or vice versa. Therefore, Granger causality provides additional evidence as to whether, and in which direction, price transmission is occurring between two series.

The hypothesis that p_1 Granger-causes p_2 and vice versa can be assessed within a Vector Autoregression (VAR) framework (see Annex) by testing the null that the coefficients of a subset of these jointly determined variables, the lagged p_1 terms, are equal to zero. In addition, Granger (1988) proposed a test for long run Granger causality within the context of the error correction representation of a cointegrated system of variables. The presence and direction of Granger causality in the long run can be assessed by testing the null that the error correction coefficients a_1 and a_2 in the VECM presented by (3) are equal to zero, a test that also reveals weak exogeneity in the econometric sense. In more detail, under $a_1 = 0$, $a_2\,^{1}\,0$, p_2 Granger-causes p_1 in the long run, under $a_2 = 0$, $a_1\,^{1}\,0$, p_1 Granger-causes p_2 in the long run, whilst under $a_1\,^{1}\,0$, $a_2\,^{1}\,0$, both series Granger-cause each other in the long run.

It is important to note that although cointegration between two price series implies Granger causality in at least one direction, the opposite is not necessarily true. In this case, as noted above in the discussion on cointegration, lack of cointegration between the two trending price series may indicate that market integration is absent, as other factors such as transaction costs determine the movements of one of the price series. However, Granger causality may exist, indicating that, although the two price series drift apart due to other factors such as non-stationary transaction costs, some price signals are passing through from one market to another. On the other hand, lack of Granger causality may not imply an absence of transmission, as price signals may be transmitted instantaneously under special circumstances. However, given the inherent dynamics of markets, we believe that this is highly unlikely.

The error correction representation also provides a framework for testing for asymmetric and nonlinear adjustment to long run equilibrium. Granger and Lee (1989) proposed an asymmetric ECM (AECM) where the speed of the adjustment of the endogenous variable depends on whether the deviation from the long run equilibrium is positive or negative. The single asymmetric ECM is specified as follows:

$$\Delta p_{1t} = \mu_1 + \alpha_1^+ \left(p_{1t-1} - \beta p_{2t-1}\right)^+ + \alpha_1^- \left(p_{1t-1} - \beta p_{2t-1}\right)^- + \sum_{i=0}^{k} \delta_i \Delta p_{2t-i} + \sum_{i=0}^{n} \gamma_i \Delta p_{1t-i} + v_{1t} \quad \text{... (5)}$$

The errors or divergences from this equilibrium are decomposed in two parts, $(p_{1t-1} - \beta p_{2t-1})^+$ and $(p_{1t-1} - \beta p_{2t-1})^-$ reflecting positive and negative disequilibria, respectively. Within this context, asymmetry occurs in the event when positive and negative divergences from the long run equilibrium between p_{1t} and p_{2t} result in changes in p_{1t} that have different magnitude. Therefore, asymmetric transmission implies that a^+_1 is not equal to a^-_1. The null of symmetry against the alternative hypothesis that adjustment is asymmetric is tested by imposing the equality restriction $a^+_1 = a^-_1$. In addition to the above, short run asymmetric transmission can also be tested by decomposing DP_{2t} in two parts reflecting price rises and price falls, and testing for equality of the corresponding short run coefficients.

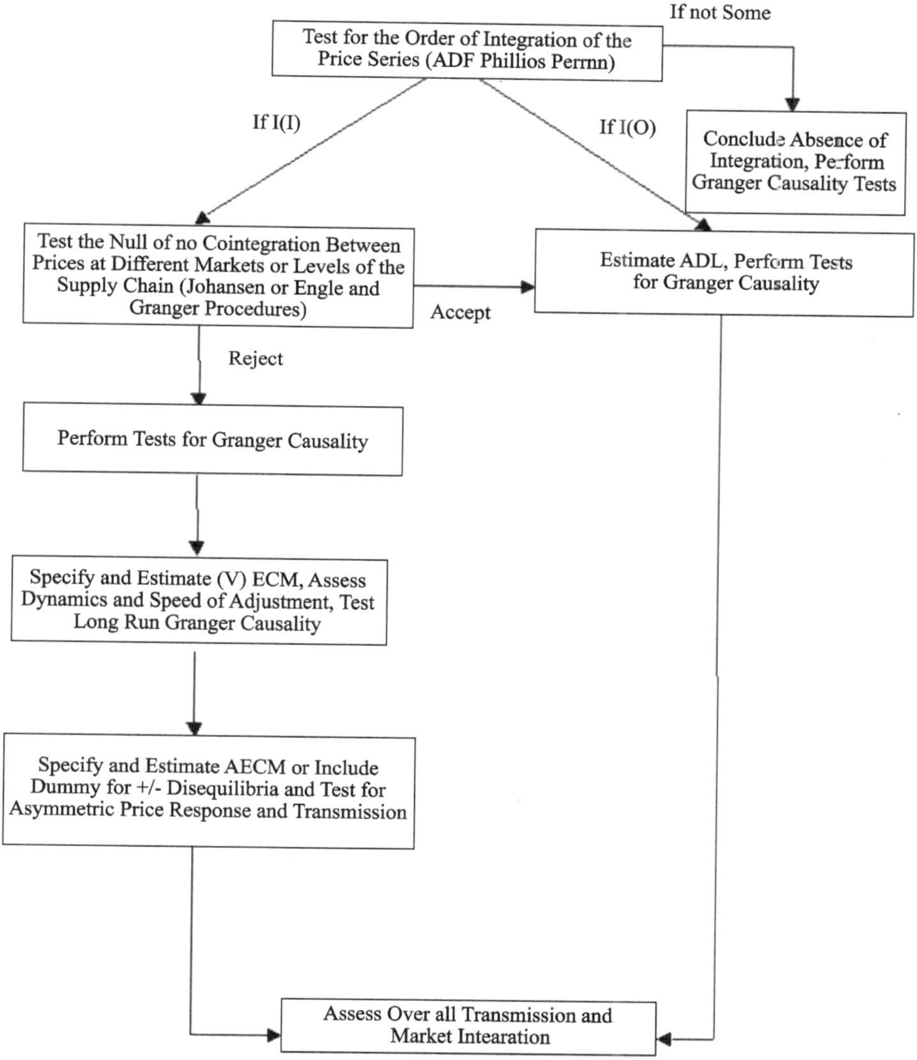

Asymmetric adjustment can be also tested by following Prakash *et al* (2001). This method involves the assignment of a dummy variable, d=0 to all the parameters of the underlying Autoregressive Distributed Lag (ADL), if there is positive disequilibrium and d=1 if there is negative disequilibrium. Asymmetric adjustment to the long run equilibrium is then tested by imposing and testing zero restrictions on the dummies' parameters.

In view of the above discussion on the empirical tools that can be used to assess the notional components of market integration and price transmission, we proceed to apply the proposed time series techniques on selected commodity markets in a sequence depicted in Diagram 1. The way in which the tests for the components of transmission have been ordered is to some extent *ad hoc*. The sequence of the tests is as follows:

i) For each pair of prices, we start by testing for the order of integration for each price utilizing the Augmented Dickey-Fuller (Dickey and Fuller, 1979) and the Phillips and Perron tests (Phillips and Perron, 1988). In the event that the series have a different order of integration, we conclude that the markets are not integrated. In the case that the series are found to be I(0), we resort to assessing the dynamics of the relationship by means of Autoregressive Distributed Lag (ADL) models. We test for Granger Causality within a Vector Autoregression (VAR) framework to assess price transmission between the markets or along the supply chain.

ii) In the event that the tests indicate that the series are integrated of the same order (say I(1)), we proceed by testing the null of non cointegration against the alternative hypothesis of one cointegrating vector using the Johansen procedure (Johansen 1988, 1991), or we test for the null of non cointegration following Engle and Granger (1987). Evidence against the null of no cointegration is taken to indicate that prices co-move and that markets are integrated. We do not impose and test for any restrictions on the cointegrating parameter estimate. As noted earlier in this section, inference on the extent of price transmission based on the size of the parameter may be misleading. In the event that the null of non cointegration is not rejected, we conclude that the markets are not integrated, and/or that we are unable to conclude that price transmission along the supply chain is complete.

iii) In the event that tests indicate that the price series are cointegrated, we proceed by focusing on the error correction representation, in the form of a (V)ECM and on examining the short run dynamics, the speed of adjustment and the direction of Granger causality in the short or the long run following Granger (1969, 1988).

iv) At the next stage, based on our results on the direction of causality, we specify AECMs and test for the null of symmetry following Granger and Lee (1989) or Prakash, Oliver and Balcombe (2001). Finally, we discuss the results and comment on the nature of price transmission and market integration.

It is important to note that the above testing framework does not identify the factors that affect market integration and price transmission. In other words, we are not able to distinguish whether price transmission and market integration is shaped by transaction costs, policy intervention that insulates the domestic markets, or by the degree of market power exerted by agents in the supply chain. For this reason, an attempt is made to complement the results with some qualitative information on the major factors that may determine the extent of transmission.

Some Policy Implications and Suggestions for Future Research

The focus here is on time series techniques to test for spatial price transmission in a number of cash and food crop markets in developing countries. The objective of the article was to provide a review of the application of time series techniques in testing market integration and to contribute to the body of research on this subject by highlighting a number of important issues.

It was stressed that although market integration, in the Enke-Samuelson-Takayama-Judge sense, and complete price transmission can be formally tested in the long run, the extent to which price signals are transmitted from one market to another is an ambiguous concept. In order to assess its extent, we decomposed the concept of price transmission into notional components: co-movement and completeness, dynamics and speed of adjustment and asymmetric response. The above definition of price transmission encompasses the case of market integration, the inherent dynamic market relationships that arise due to inertia or discontinuities in trade, as well as non linearities that may arise due to distortions in arbitrage. More importantly, it implies hypotheses, through its components, that are testable within a cointegration - error correction model framework. The testing framework was applied to a number of cash and food crop markets in developing countries. In general, given that the cointegrating parameter does not reflect the "elasticity of price transmission" well, we proposed that the assessment of the extent of market integration and price transmission should be based on ECMs with both symmetric and asymmetric adjustment.

Fisheries policies may or may not hinder market integration, depending on the nature of the policy instruments employed. In general, for markets that are subject to policies, the speed of adjustment, as reflected by the error correction coefficients, was estimated to be relatively low. Although several authors stress that policies impede the extent of price transmission (see for example Mundlak and Larson, 1992; Quiroz and Soto, 1996; Baffes and Ajwad, 2001; Abdulai, 2000; Sharma, 2002), it should be noted that other reasons such as high transaction costs and other distortions may also be the cause for slow adjustment.

Non linearities and asymmetric adjustment remain an important issue to be explored especially when the objective of the research is to provide a price transmission mechanism that can be incorporated in a structural partial equilibrium model. Although asymmetric adjustment may also be the outcome of market imperfections, it is plausible that price support policies result in positive and negative changes in the international price affecting the domestic market in different ways. More importantly, such policies may imply a "threshold" or a minimum price, above which, transmission of price signals take place. Such a discrete adjustment process implies that movements towards a long run equilibrium do not take place at all points in time, but only when the divergence from equilibrium exceeds a certain threshold. For example, policies such as price support mechanisms and tariff rate quotas may result in such an adjustment process. In the former case, governments may intervene in the market when market prices fall below a floor level, whilst in the latter, international price signals pass-through when import volumes are sufficiently within, or out of the quota. Thus, future research may focus on two-regime threshold cointegration, which may be beneficial, as it provides additional information in the form of the threshold, if the objective of the analysis is the development of price transmission mechanisms for structural models.

Apart from assessing the effect of seafood and trade policies on market integration, threshold cointegration applied in commodity markets of developing countries may also provide a rough indication of transfer costs. Transfer costs in developing country markets may give rise to a threshold over which arbitrage possibilities are obliterated, resulting in an absence of market integration. Thus, a threshold cointegration framework can encompass the possibility of non-stationary transfer costs and provide valuable information that can lead to policy prescriptions.

References

George Rapsomanikis, David Hallam and Piero Conforti, 2003. Market Integration and Price Transmission in Selected Food and Cash Crop Markets of Developing Countries: Review and Applications, in Commodity Market Review, Food and Agricultural Organisation (FAO), Rome Italy.

Singh, Naorem Dinesh, 2018. Propagative Causal Price Transmission in Indian Shrimp Exports Markets, Unpublished Ph.D. thesis, Fisheries Economics, Extension and Statistics Division, ICAR- Central Institute of Fisheries Education, Mumbai.

14

Big Data in Fisheries

Ravichandran S.

Introduction

Indian agriculture, including fisheries, is highly monsoon dependent and effect of climate change is immense on its large tracts of rainfed farms. The effects of climate change could be thwarted through better adaptation techniques using all information available with various agencies concerned with agriculture and using advanced analytical tools. Presently, in the country, there is a gap in the capability in capturing, processing and analyzing huge source of information generated in Indian agriculture. This task can be accomplished using Big Data Platform which works in cluster mode and provides the capability of handling huge volume of data, capturing streaming data or real time data and also can accommodate unstructured data. As per FAO, 70 percent in increase in food production in another 30 years is required to overcome the increasing global food demand. It may not be possible to achieve that target due to variety of problems such as reduction in land for cultivation, climate change, and various other factors, such as increasing the fish production etc. Improvement in crop genetics, application of fertilizer in higher dosages, irrigation facilities, etc. may not be possible in future. We need to create a big data platform involving important parameters on crop production viz., soil, soil temperature, water requirements, and relative humidity, pests and diseases, and micro and macro nutrients in temporal and spatial. Now, since available data is properly utilized for making valid analytics by the farming community due to certain issues. Even the data created is incompatible, data available from various sources is also not compatible among themselves due to which decision making is also a big challenge. Due to the incompatibility of agricultural data from different sources, farmers do not have the accessibility to vast pool of available information for making effective, efficient and timely decisions.

Big Data in Agriculture (including fisheries)

Big data, an emerging data driven system helps in precision agriculture based on data collected. This helps farmers in making efficient decisions on sowing, planting, fertilizing and various other farm operations including harvesting crops at the appropriate time. Of late, sensors are placed in the fields to collect data on related parameters viz., soil temperature and moisture relative humidity, wind speed, weather, etc. which help in crop growth. Images obtained using satellites are helpful in monitoring the crop growth during crop period. Agro-produce includes not only yields obtained from crops but also include produce from horticulture, livestock, fisheries, and other systems also. Agriculture is dependent on various parameters such as climate, seed, soil cultivation practices, agronomic practices, irrigation requirement, fertilizers pesticides, weeds, harvesting, post harvesting techniques, etc. Data collected on all these parameters would help in estimation of yield and other relevant parameters. Governments, Universities, Research departments, Agro-business and Agri-input companies generate, maintain and use voluminous data related to various farm management practices and other related issues say insurance marketing, supply chain, packaging, distribution, etc. This voluminous and wide variety of data poses several challenges for making effective decisions. Information on various fish related parameters can be collected from the stakeholders and a big data platform can be developed for effective decision making.

Related Literature

In literature, few works on data analytics have been reported over a period of time. These applications though have not been related to big data as such but they all deal with data collection and subsequent analysis based on which various important parameters have been obtained. Average yields from large tracts obtained temporally were utilized to obtain trends in yields of certain crops (Thompson, 1969; Gage and Safir 2011). It was also observed that an increasing trend was noticed for the past several decades. This was mainly due to technological advancements in genetic and crop managements especially due to over usage of chemical fertilizers. It was observed that the yield deviations were directly related to monthly average data on certain weather parameters. Thompson (1986) developed a statistical model to assess the effect of climate and weather on corn production in five Midwestern states in USA. It was found out that the precipitation during September-June, June temperature and temperature and rainfall in July and August were closely related with yield variations in corn crop. Gage and Safir (2011) made use of climate and stress indices of crops for working out regional yield trend. It was noticed that this particular method significantly improved historical predictions of yield in corn and soybean. Lobell *et al.* (2011, 2013) utilized various statistical models by involving temperature and maize yield. It was found out there is significant yield reduction due to increase in temperature. It was also found out that increase in temperature played a greater role in yield decrease. As a note of precaution, models developed for a particular dataset cannot be extended to other datasets involving space and time because of certain factors such as soil types, landscapes, and weather. Models basically are data dependent and hence the crop-weather models developed for a certain situation cannot be applied to all the areas uniformly. Effects of changes in agricultural technology have to be taken into consideration while simulating yield using various statistical models. Hence, the problem associated with statistical models in respect crops is that the forecasts may be made outside the range of the data when weather and technology information was utilized for which the model was developed. Statistical models will help in knowing about past yields and future forecasting can also be carried out depending upon the appropriateness and efficiency of the models (Gage and Safir, 2011; Lobell *et al.*, 2011). Big Data's biggest strength is its versatility and application in many industries. It has also enormous impact on many areas of daily life. Application of Big Data can be a game changer in many applications. A lot of data is generated in all walks of life and especially in many industries and these data have to be analyzed properly depending upon the type of data and meaningful information has to be brought that will help the stakeholders.

Big data analytical tools will certainly revolutionize agriculture in the years to come. Farm Link App focuses on the data quality while aggregating and analyzing information it uses. In 2011, combine-leasing company made use of sensors and computers in their experimental fields and calculated yield from 150 square feet of farmers' fields. So far, data consisting of more than a billion records have been collected across 26 states of US on 50 variables related to yield, such as soil type, slope, elevation, sunlight, rain, and temperature. Big data analytical tool called True Harvest was rolled out in 2014 using the information collected. This tool is very handy to farmers to evaluate on water, fertilizer, and pesticide application, etc. for effective farm management decisions. Manuel Mendoza, Maria G. Pennino and Jose M.Bellido (2012) dealt with tree-Based machine learning analysis for fisheries research. DOLFIN, a business intelligence platform is being developed.

Big Data in Fisheries

Big data offers vital tools for managing fisheries. For example, fishing trips could be tailored to meet quotas in the shortest time frame, reducing fuel consumption and crew costs. Real-time catch data could help administrations close a fishing zone immediately once quota is reached, rather than weeks later using paper logbooks and hours of human analysis. Fishing ministries could use big data analytics to move fast to preserve marine protected areas and their fish stocks. Data-mining techniques can be used to identify the most interesting correlations among vessel data, catch data, Vessel Management System (VMS), logbooks, sensor data and oceanographic data. Dashboard can be prepared to search by zone, species type, vessel and period. Fleet strategy can be optimized using data mining. Captains can decide where to go before leaving the harbour, optimizing the time spent to reach quotas. Fisheries administration can monitor catch efforts over an entire region and decide which zones to close, improving management of stocks and fishing licenses. In addition, some governments give fuel allowances, which could be better managed and distributed according to vessel's behavior. The insights provided by big data and advanced analytics offer a valuable way forward for both regulators and fishermen, enabling them to capitalize on their data-an asset as valuable as technology. Fisheries information needs more attention as the transition not only to a more sustainable blue economy but also to the knowledge economy. Neglecting data assets means losing efficiency and resources. Adopting advanced data analytics is the key to succeeding in the future of fisheries. Advanced analytics may help struggling fisheries thrive while simultaneously protecting endangered ocean resources.

Big Data for India

Agriculture plays a vital role in Indian economy. More than 60 per cent of the population is directly or indirectly dependent on agriculture. Indian agriculture is highly monsoon dependent and effect of climate change is immense on its large tracts of rain fed farms. Natural resources like water and soil were getting increasingly overexploitation leading to pollution. Similarly, the excessive use of fertilizers and pesticides/herbicides has led to problems like ground water pollution and resurgence of pest resistance. The effects of climate change could be thwarted through better adaptation techniques using maximum information from all the agencies concerned with the agriculture and using advanced state-of-the-art analytical tools to draw useful inferences. Agriculture generates high volume of data in real time and in a highly unstructured format. Presently, in the country, there is a gap in the capability in capturing processing and analyzing this huge source of information generated in Indian agriculture such as water, inputs, outputs, weather, market, etc. This task can be accomplished using Big Data Platform which has three components, namely, volume velocity, and variety. The big data platform works in cluster mode and it does away the requirement of hosting huge capacity servers and also do away with frequent replacements. It has the capability of handling huge volume of data, can capture streaming data or real time data and also can accommodate unstructured data.

Presently, the crop advisory system is through line department and extension officials of the state government which is basically using eye-appraisal. It is not able to reach all the farmers or stakeholders due to shortage of manpower and not using latest ICT tools. Even though many organizations are involved in crop advisory services, they are not integrated into one platform and hence, the data is not effective utilized for effective decision making. In this direction, we are planning to develop a common platform for pooling and effective utilization of information related to agriculture collected by various functionaries and agencies. Other stakeholders who are involved in various advisory services can also be benefitted through joining the common platform and thereby sharing their repository which will enable better advisory services. Customized advisory services can also be provided depending upon the requirements of the stakeholders. Advanced crop planning, forecasting, and market based decision support systems will help in conserving natural resources like soil, water, etc., sustainable food production, and improving the efficiency of advisory services of various agencies.

Global Agricultural Research Data Innovation & Acceleration Network (GARDIAN) – CGIAR's Big Data Platform

The platform involves 15 Research Centers and 15 Research Programs and Platforms, and more than 8,000 researchers within CGIR as well as thousands of partners in Government and the private sector. Encompasses one hundred thousand publications (96175) and over Two thousand and five hundred data sets (2852) which could be discovered, accessed and reused. This is the first solid step towards making all the scientific knowledge the CGIAR network generates openly discoverable, accessible and usable. GARDIAN is becoming the Google search of all CGIAR knowledge which makes us to discover everything from the historical knowledge residing in legacy systems to the new scientific discoveries that the flagship research programmes make. Whatever the CGIAR network does, it is discoverable and accessible irrespective of scientists, projects, publications, data sets, models and software. Everything is appropriately described and classified, at the world's fingertips. GARDIAN is becoming the data science brain that will power new discoveries. Also, powering a novel way to communicate science to the world.

ICAR Big Data

ICAR is in process of developing a big data platform for achieving some of the objectives viz., 1. To develop big data based repository system on various agricultural components. 2.To develop dynamic crop planning and forecasting models integrating crop growth, agro meteorological, environment, and market dynamics and to develop crop advisory systems. 3. To create centre of excellence in big data analytics in agriculture for capacity building of researchers involved in crop planning and farm advisory services in India. These objectives can be achieved by randomly selecting fewer states and in fewer fields of research stations of Indian Council of Agricultural Research/ Krishi Vigyan Kendras. High resolution sensors can be installed in these fields. Data on crop, soil, weather, pest incidence, etc. will be obtained from the sensors and the data will be stored in the computer systems connected to the sensors which records the information with high frequency. The data thus stored in these systems in different places will form the cluster and big data analytics could be carried out with the help of various big data analytical tools like Hadoop, Hive, Spark, Sqoop, RHadoop, etc. The big data framework brings all stakeholders who works/related to Indian agriculture together and enhances their capacity and knowledge base for providing advisory services to farmers and other stakeholders. Huge repository creation on which advance research can be taken up for better/ improved pattern/ forecast. Consistency in creating structured data and ease in retrieval of quality and needful/exhaustive information for better analysis or decision making. Using real time data, dynamic advisories systems can be developed which will enhance the income of stakeholder mitigate the risks associated with farming at much lower cost. Creation of a pool of trained man power in handling big data related to agriculture. Analytical techniques such as optimization of cropping pattern, planting date determination, crop phenological models, dynamic pest and disease incidence models, machine learning models, simulation models, and market based price forecast and decision models will be applied to accomplish our objectives. Hence big data platform can help in creating a system through which can generate accurate forecasts of crop growth, provides information about timing of sowing and harvest, indicates the level of pest and disease incidence and alerts when it crosses the threshold level, and also provides useful alerts about market information. It helps policy makers and agricultural agencies in designing incentive systems which entices farmers to adopt climate smart agricultural practices. Big data analytics in agriculture applications will help in weather forecast, achieving higher productivity, reduction in harvesting and efficient fertilizer and pesticide usage. It is also pertinent to infer that big data obtained from different sources using ICT components for precision agriculture and structured and unstructured data calls for application of big data tools.

KRISHI – ICAR Efforts

Knowledge based Resources Information Systems Hub for Innovations in Agriculture (KRISHI), is an initiative of ICAR to bring its knowledge resources to all stakeholders at one place. The portal is being developed as a centralized data repository system of ICAR consisting of Technology, data generated through experiments, surveys, observational studies, geo-spatial data, publications, learning resources etc.

Conclusion

Through Big data, scientists and researchers involved will be able to pool all the related information a farmer requires for efficient farming. On the whole, big data in agriculture will bring together all necessary data on various factors viz., biological, environmental, and human factors used for crop growth by providing farming community with accurate and up-to-date information on growing crops more efficiently and economically. While human knowledge and technological innovations and advancements helped humanity to grow for millions of years, advancements in computing helps the scientific community to achieve breakthroughs in collection, exchange of data followed by processing, and effective decision making in transforming farming into a sustainable and profitable enterprise so that the ever growing in food demand of 9 billion mouths by 2050 be met with. Efforts should be made by the stakeholders in fisheries, including researchers, should plan for a big data platform for fisheries sector so that all information pertaining to fisheries related data would be available in one particular place so that decision making would be easier for planners.

15

Limited Dependent Variable Models

Kiresur V.R.

Introduction

The influence of various socio-economic factors on the willingness of the decision makers to adopt new technologies has been investigated by a number of studies (Roe, 1983; Shakya and Flinn, 1985; Thomas *et al.,* 1990). In most studies of adoption behaviour, the dependent variable is constrained to lie between 0 and 1 and the models used are exponential functions (Kebede *et al.,* 1990). However, the decision to adopt a new technology can be very effectively captured using binary choice models. Binary choice models are appropriate when the choice between two alternatives depends on the characteristics of the problem. Application of a linear probability model to this type of problem, however, suffers from a number of deficiencies (Capps and Kramer, 1985), particularly, the one associated with the estimated probabilities in some cases being greater than one or lesser than zero as a result of neglecting significant interaction effects (Mingche, 1977). These deficiencies could be circumvented through the use of a monotonic transformation (probit or logit specification) which guaranties that predictions lie within the unit interval (Capps and Kramer, 1985). Univariate logit and probit models and their modified forms have been used extensively to study the adoption behaviour of farmers and consumers (Schmidt and Strauss, 1975; Garcia *et al.,* 1983; Shakya and Flinn, 1985; Harper *et al.,* 1990). According to Hanushek and Jackson (1977), the choice between logit and probit models is largely a matter of convenience. However, Maddala (1983) and Shakya and Flinn (1985) have recommended probit models for functional forms with limited dependent variables that are continuous between 0 and 1, and logit models for discrete dependent variables.

Model Specification

Univariate logit model is generally used to study the adoption behaviour of farmers and consumers.

The model is estimated using the maximum likelihood method.

$\ln [\{P(m/X)\} / \{P(t/X)\}] = XB + E$ ... (1)

or $\ln [\{P(m/X)\} / \{1 \; P(m/X)\}] = XB + E$ ... (2)

where,

X = vector of explanatory variables;

P(m/X) = probability of an individual farmer adopting MT, given X;

P(t/X) = 1 P(m/X)

= probability of an individual farmer adopting the TT (or not adopting the MT) given the level of X;

$[\{P(m/X)\} / \{P(t/X)\}] = [\{P(m/X)\} / \{1 \; P(m/X)\}]$ = the relative odds of adopting versus not adopting the MT;

B = Vector of response coefficients; and

E = Vector of random disturbances.

Case Study-Results

Groundwater Marketing in Karnataka: Key Issues in Sustainability (Nayak and Kiresur, 2007)

Maximum Likelihood Estimates of the Logit Model for a farmer's decision to participate in groundwater market.

The study mainly uses primary data obtained from 120 farmers spread across Belgaum, Bijapur and Bagalkot districts of Karnataka. Data were processed using tabular analysis and Logit Model. Estimated logit model was a good fit as indicated by the per cent-correct classification (90.90 per cent) of the binary (0,1) dependent variable. The proportion of the participants who were predicted correctly was to the tune of 94 per cent whereas that of non-participants was 86.8 per cent. Only five out of 16 variables included in the model were significant. Three variables (NWW, LHD and IRRIGFUR) were significant at 1 per cent probability level, while one (AREAKH) was significant at 5 per cent level and AGE at 10 per cent level, respectively. AGE, NWW and IRRIGFUR were found to negatively influence the farmer's decision to participate in water market. On the other hand, LHD and AREAKH positively and significantly influenced the probability of a farmer's decision to participate in water market. To interpret the effect of change in the values of explanatory variables on the probability of a farmer's decision to participate in water market, the change in probability (ΔP, the derivative at mean) was calculated. For every one year increase the age of the farmer, the probability of a farmer's decision to participate in water market decreases by 3.57 per cent. Similarly, a unit increase in IRRIGFUR decreased the probability of a farmer's decision to participate in water market by 29.91 per cent. Farmers suggested changing the irrigation method from furrow to drip/sprinkler for conserving water. Annual crops fetched highest average gross returns per ha for both sellers and buyers. Well-owners sold water to neighborhood farmers only if they had surplus water. The most serious negative externalities of groundwater marketing were reduction of water outflow during pumping and fall in water table. Some policy implications that emerged from the study were implementation of effective groundwater recharging technologies, monitoring of prescribed inter-well distance, nationalisation of groundwater resource and establishment of 'Community Borewells for Irrigation', among others.

Discriminant Function

A statistical tool used to discriminate between two or more classes of certain relevant characters. Multivariate problem is reduced to a simple uni-variate problem by assigning appropriate weighting coefficients. The single weighted composite score will minimize error of misclassifying an individual unit from one group to the other. Linear Discriminant Function is a tool used to discriminate between (only) TWO classes of individuals or objects, for example, classifying borrowers into defaulters and non-defaulters, farmers as adopters and non-adopters of a technology, etc.

The linear form of the function:

$$Z = \sum_{i=1}^{p} b_i X_i \qquad \text{... (1)}$$

where,

Z = total discriminant score for the two groups,

xi's = variables or factors or characteristics selected to discriminate the groups,

bi's = linear discriminant coefficients,

p = the number of characteristics

Weighting coefficients are to be derived in such a way that the 't' statistic or 'F' ratio between groups will be maximum (Pandey and Muralidharan, 1977). The bi's in equation (1) can be estimated by multiplying the inverse of the covariance matrix by the vector of the mean differences as follows:

$$b = S^{-1} d \qquad \text{... (2)}$$

where,

'S' is the variance-covariance matrix;

'd' is the vector of difference of mean values of 'p' characters for the two groups; and
'b' is the vector of the coefficients of discriminant function

The Mahalnobis 'D2' statistics is used to measure the distance between the two groups and it is given by:

$$D2 = \sum_{i=1}^{P} bi\ di \qquad \text{... (3)}$$

After transformation, this 'D2' statistics becomes an 'F' Statistic of the following form, which is then used to test whether the two groups are significantly different from each other.

$$F = \frac{n1\,n2(n1 + n2 - p - 1)}{P(n1 + n2)(n1 + n2 - 2)} D2 \qquad \text{... (4)}$$

where,

n1 = number of sample units in the first group,

n2 = number of sample units in the second group,

p = the number of characteristics / factors,

D2 = Mahalanobis Statistic.

'F' is tested for its significance with p and (n1 + n2-p-1) degrees of freedom.

If 'F' is significant, then it implies that the function is capable of discriminating the two groups and is ready for classificatory purpose. Then, we may proceed to compute the score values of Z1 and Z2 as follows:

$$Z1 = \sum_{i=1}^{P} bi\ Xi1\ (\text{for group} - 1) \qquad \text{... (5)}$$

$$Z2 = \sum_{i=1}^{P} bi\ Xi2\ (\text{for group} - 2)$$

Next, the cutting point or discriminatory point ZC for classifying the individuals into two groups is calculated as:

$$Zc = \frac{Z1 + Z2}{2} \qquad \text{... (6)}$$

For an individual sample unit, the Z value can be calculated by:

$$Z = \sum_{i=1}^{P} biXi \qquad \text{... (7)}$$

If the individual 'Z' value is more than the Zc value, the individual is classified into group-2 and if it is less than Zc value, the individual is put under group-1. In a similar manner, the probability of mis-classifying an individual unit from one group to other could be obtained by computing the Ze score of Zc as follows:

$$Ze(1) = \frac{Zc - Z1}{\ddot{O}V(Z)} \quad \text{... (8)}$$

$$Ze(2) = \frac{Zc - Z2}{\ddot{O}V(Z)}$$

Since the cutting point Zc was calculated to lie equidistant between Z1 and Z2, the two Ze score values are identical except for the sign.

Application of Discriminant Function

The constraints hindering adoption of HYVs of rice in Puri district of Orissa State was estimated (Shanti Sarup and R. K. Pandey (1982). The sample farmers (113) were classified into two groups; the farmers growing HYVs in full or in part were grouped as 'A' (55) and those not growing at all as 'B' (58). The information on 13 characteristics was used for discriminating between the groups.

Form of Discriminant Function:

$$Z = L1X1 + L2X2 + \ldots\ldots\ldots\ldots L13\,X13 = \underset{i=1}{S}\ Li\,Xi$$

where,

Z = the total discriminating score for adopters and non-adopters,

X1 = average yield at the farm (in kg/acre) in the previous year,

X2 = total area cultivated at the farm (in acres) by the farmer,

X3 = area leased-in (in acres),

X4 = number of fragments of the operated area,

X5 = fertilizer application (if yes=1, otherwise=0),

X6 = plant protection measures (if taken=1, otherwise=0),

X7 = number of family members,

X8 = number of literates in the family,

X9 = number of family members engaged in agriculture,

X10 = any additional source of income other than crop production (if yes=1, otherwise=0)

X11 = amount of loan taken during the current season (₹),

X12 = total outstanding amount of loan (₹),

X13 = type of land; if more than 50 per cent of owned is low=0, otherwise=1

Li = (i=1,2,........13) are the coefficients of linear discriminant function.

Case Study – Results

The discriminant function thus estimated was:

$$Z = 0.0020\ X1 + 0.03900\ X2 + 0.49000\ X3 + 0.08750\ X4 + 2.43900\ X5 + 0.71460\ X6 - 0.22590\ X7 + 0.29850\ X8 + 0.12090\ X9 + 0.09480\ X10 - 0.00070\ X11 + 0.00003 X12 + 0.83630\ X13$$

D2 = 2.33

F = 4.52*** (d.f.: 13,99)

The average yield at the farm (X1), area under cultivation (X2), fertilizer application (X5), plant protection measures taken (X6), number of literates in the family (X8) and type of land (X13) are the major characteristics which classified the farmers into the groups of those growing and not growing HYVs.

Their respective discriminating powers are 11.05, 2.28, 48.90, 10.73, 12.85 and 5.49 per cent.

References

Eric A. Hanushek, John E. Jackson, 1977. Statistical methods for social scientists, 374 pp.

Li, Mingche M, 1977. A logit model of homeownership, Econometrica, 45(5), 1081-1097.

Oral Capps Jr., Randall A. Kramer, 1985. Analysis of food stamp participation using qualitative choice models, American Journal of Agricultural Economics, 67(1), 49-59.

Padma B Shakya and J.C. Flinn, 2008. Adoption of modern varieties and fertilizer use on rice in the eastern Tarai of Nepal, Journal of Agricultural Economics, 36(3): 409-419 (DOI:10.1111/j.1477-9552.1985.tb00188.x).

Peter Schmidt and Robert Strauss, 1975. The prediction of occupation using multiple logit models, International Economic Review, 16, (2), 471-486.

Yohannes Kebede, Kisan Gunjal, Garth Coffin, 1990. Adoption of new technologies in Ethiopian agriculture: The case of Tegulet-Bulga district Shoa province, Agricultural Economics, 4(1), 27-43.

16

Variance Components and Estimation of Genetic Parameters

Shrinivas Jahageerdar

Biometrical Genetics - Historical Background

It is a science built on the principles of Mendelian genetics and the way in which the genetic factors influence the statistical properties of traits in families and populations. The roots of biometrical genetics go back to Gregor Mendel, who first demonstrated the existence of the hereditary factors and to Francis Galton who first applied statistical methods to study biological inheritance. Galton introduced the concepts of regression and correlation for studying the continuous variation in humans, but was not aware of the principles of genetics discovered by Mendel. Galton, Karl Pearson and Walter Weldon together started the journal Biometrika in 1901, with the aim of promoting the statistical studies in biological phenomena. The differences of opinion persisted between the Biometricians and Mendelians in the early 20th Century. In 1918 Ronald A. Fisher published a landmark paper 'The correlation between relatives on the supposition of Mendelian inheritance, in which he showed the two seemingly contradictory approaches could be reconciled in terms of a simple biometrical genetic model. Ronald A. Fisher's paper laid the foundation for the discipline of Biometrical Genetics.

Foundation

a) Mendelian Inheritance

Mendel proposed three principles based on his work and now known as Mendel's laws of inheritance

- ***Law of uniformity:*** if two homozygote parents of different alleles cross together, all offspring must have the same genotype
- ***Law of segregation:*** an individual receives with the equal probability one of the two genes from the genotype of the mother and similarly from that of the father.

- ***Law of independent assortment:*** the segregation of the genes for one trait is independent of the segregation of genes for other traits in the formation of reproductive cells

In association studies, the law of segregation is the basis for the transmission disequilibrium test (TDT). The possible violation of law of independent assortment forms the basis of the gene mapping technique, linkage analysis. These principles are used to check the quality of parent-offspring genetic data. Mendelian principles though developed for dichotomous traits, they can be applied to the inheritance of traits influenced by multiple genes and are fundamental to genetic analysis.

b) Random Mating

Many aspects of biometrical genetics implicitly assume that the mating within the population will occur at random, that is, an individual's genotype at a given locus will not influence mate choice. Random mating with respect to a particular locus results in a characteristic relationship between the allele frequencies in the mating pairs and the alleles as well as the genotype frequencies in the offspring sample.

c) Hardy-Weinberg Equilibrium

In a large random mating population, in the absence of mutation, migration, and selection, the gene frequency and genotype frequency remain constant generation after generation, and there exist a relationship between gene frequency and genotype frequency in that the genotype frequency is square of the gene frequency.

Frequency of Allele A1=p=P+½H

Frequency of Allele A2=1=Q+½H

There are two implications of HWE:

- Allele frequencies are unchanged from one generation to the next
- The genotype frequencies in the offspring can be directly predicted from the allele frequencies in the parental generation

Distortion of HWE is referred as Hardy-Weinberg Disequilibrium. HW disequilibrium results due to violation of assumptions or due to genotyping error. HWE is a necessary assumption for haplotype phasing algorithms. It is a key component of data quality checks for large-scale genotyping. Major deviation of HWE may be a sign of association, but current genotyping technologies can cause massive deviations through systemic error, such as heterozygote dropout.

Polygenic Inheritance- Polygenic Model for Quantitative Traits

Continuous variation can be explained in terms of genetic differences if we postulate that not one, but many genetic loci (i.e., polygenes) are involved in determining the quantitative trait. The polygenic model was proposed by Fisher based on the central limit theorem, which states that the '*sum of many independent random variables tends to be normally distributed, that is acquires the familiar bell-shaped appearances in its frequency distribution'*. From genetics point, this translates to the aggregation of many genes of individually small effect yielding the continuous variation observed for most quantitative traits.

This conclusion is crucial as it provides the explanation for how Mendelian genetics of inheritance can be responsible for continuous nature of individual differences in many biological traits. Quantitative traits are also called as multifactorial or complex traits, so as to allow for environmental influences on traits with multiple genetic factors. The complexity of the traits can increase even further by considering interactions between multiple genes (Epistasis) and Genotype- Environment (G*E) interactions.

Partitioning of Variance

The mean-centered genotypic value (G) can be partitioned in to the additive genetic effect (A) and dominance effect (D)

$G = A + D$

The component of genetic variance σ^2G can be partitioned in to

$\sigma^2G = \sigma^2A + \sigma^2D + 2cov(AD)$

where, σ^2A and σ^2D are the additive and dominance genetic variances, respectively

Usually it is assumed that the cov (AD) = 0. Estimating the σ^2A is equivalent to the estimation of the breeding values (BVs). Generally speaking, the additive genetic variance σ^2A defines the regression variance in a standard regression model, while the dominance genetic variance σ^2D is the residual variance.

Fixed Effects Models- Example 1- (Fish varieties)

Consider a farmer carrying out a small experiment in the backyard farm with 24 fish, 6 fish of each of 4 varieties that the farmer is particularly interested in, through having tried them occasionally in recent times. Comparison of the four varieties is now to be made in the 12' x 8' pond space available. Each fish is allocated randomly to one of the 2' x 2' squares cages. If $y_{i'}$, is the growth of individual fish j from of variety i (for i = 1,.., 4 and j = 1 ,..., 6), a possible model for y_{ij} would be

$E(Y_{ij}) = \mu_i$ ... (1)

where, E represents expectation and μ_i is the expected growth from a fish variety i

If we wanted to write $\mu_i = \mu + \alpha_i$ we would then have

$E(Y_{ij}) = \mu + \alpha_i$ ... (2)

where, μ is a general mean and α_i is the effect on growth of fish due to the fish being variety i.

In this modeling of the expected value of y_{ij} each μ_i, (or μ and each α_i) is considered as a fixed unknown constant, the magnitudes of which we wish, in some general sense, to estimate; e.g., we might want to estimate μ_i, and $\mu + \alpha_i$, or $\mu_1 - \mu_2$. In doing this, the μ_{ij}, (or the α_is) correspond to the four different varieties that the farmer is interested in. They are four very specific varieties of interest, and in using them the farmer has no thought for any other varieties. This is the concept of fixed effects. Attention is fixed upon just the varieties in the experiment, upon these and no others, and so the effects are called fixed effects. And because all the effects in (2) are fixed effects, the model is called a fixed effects model. It is also called Model I, so named by Eisenhart (1947).

Armed with $E(Y_{ij}) = \mu + \alpha_i$ (2), we now define the deviation of $\mathbf{y_{ij}}$, from its expected value $E(y_{ij},)$ as residual error:

$e_{ij} = y_{ij} - E(y_{ij}) = (y_{ij} - (\mu + \alpha_i))$ ... (3)

This gives $y_{ij} = \mu + \alpha_i + e_{ij}$ or equivalently $y_{ij} = \mu_i + e_{ij}$

A consequence of the definition of the residual e_{ij}, is that it is a random variable with mean zero.

$E(e_{ij},) = E[y_{ij}, - E(y_{ij})] = 0$ ... (4)

But that definition carries no consequences so far as second moments are concerned. Therefore, we attribute a variance and covariance structure to the e_{ij}:

- first, that every e_{ij}, has the same variance, σ^2
- and second, that the e_{ij}s are independently and identically distributed and that pairs of different $\mathbf{e_{ij}}$s have zero covariance.

Thus, using var (.) to denote variance and cov (., .) for covariance

var $(e_{ij}) = \sigma^2 e$, for all values of i and j ... (5)

cov $(e_{ij}\, e_{i'j'}) = 0$ except for i = i' and j = j' ... (6)

In light of equation (4) [$E(e_{ij},) = E[y_{ij}, - E(y_{ij})] = 0$. This means that

$\text{var}(e_{ij}) = \sigma^2 e = \underset{ij}{E}(e^2)$

$E(e_{ij}\, e_{i'j'}) = 0$ *except for* $i = i'$ *and* $j = j'$ ... (7)

The manner in which data are obtained always affects inferences that can be drawn from them. We therefore describe a sampling process pertinent to this fixed effects model. The data are envisaged as being one possible set of data involving the same fish varieties that could be derived from repetitions of the experiment, repetitions for each of which a different sample of 6 fish of each variety would be used. This would lead on each occasion to a set of **es** that would be a random sample from a population of error terms having zero mean, variance σ^2e and zero covariances. It is the probability distribution associated with the es that provides the means for making inferences about functions of the μ_is (or of μ and the α_is) and about σ^2e.

The all-important feature of fixed effects is that they are deemed to be constants representing the effects on the response variable **y** of the different levels of the factor concerned, in this case the varieties of fish. These varieties are the levels of the factor of particular interest, chosen because of interest in those varieties in the experiment. But they could just as well be different fertilizers applied to ponds, different culture systems in the same region, different machines used in a manufacturing process, different drugs given for the same illness, and so on. The possibilities are legion, as are the varieties of models and their complexities, reaching far beyond those of (1)-(7).

Random Effects Models: Example 1

Suppose a new form of injectable GnRH is being tested using 10 different hatcheries in Tamil Nadu State. It is not unreasonable to think of those hatcheries as a randomly chosen sample of hatcheries from a population of hatcheries (i.e., breeders who administer the injections). If hatchery i has n_i fish in the trial and the measured response of fish j in hatchery i is y_{ij} then a possible model would be

$E(y_{ij}) = \mu + \alpha_{ij}$ ... (9)

Although (9) is algebraically the same as ($E(Y_{ij}) = \mu + \alpha_i$ ---2) for Examples 1 and 2, some assumptions underlying it are different. In Example 1 each α_i is a fixed effect, the effect on growth of the fish having belonged to a specific variety **i**, a treatment that is a pre-decided variety of interest. But in (9) α_i is the effect on spawning level of the observed fish having been injected in hatchery i; and hatchery i is just one hatchery, that one from among the randomly chosen hatcheries that happened to be numbered i in the trial. Since the hatcheries have been chosen randomly with the object of treating them as a representation of the population of all hatcheries in TN State, and from which inferences can and will be made about that population. The one labeled **i** is of no particular interest of itself to the trial; it is of interest solely as being one of the 10 hatcheries randomly chosen from a larger population of hatcheries. This is a characteristic of random effects: they can be used as the basis for making inferences about populations from which they have come.

Random Effects Models

Thus α_i is a random effect; as such it is, indeed, a random variable. More precisely, α_i corresponding to the hatchery that has been assigned label i is the (unknowable) realization of a random variable "hatchery effect" appropriate to that hatchery labeled i. However, for notational convenience we judiciously ignore the distinction between a random variable and a realized value of it and let α_i do double duty for both. With the α_i s being treated as random variables, we must attribute probability properties to them.

There are two that are customarily employed: first, that all α_i are independently and identically distributed (i.i.d.); second, that they have zero mean, and then, that they all have the same variance, $\sigma^2\alpha$. We summarize this as

$\alpha i \sim$ i.i.d $(0, \sigma^2\alpha)$ for all I

Consequences of this are

$E(\alpha i)= 0$ for all i

$$\text{Var } \alpha_i = E\left[\alpha_i - E(\alpha_i)\right]^2 = \underset{i}{E}\left(\alpha^2\right) = \underset{i}{\sigma}^2 \alpha$$

A second outcome of treating the **αs** as random variables is that we must consider with more forethought, because it is really a conditional mean.

$E(y_{ij})=\mu + \alpha_i$ ---of 9

Suppose for the moment that α^* represents the random variable "hatchery effect", and that for the hatchery labeled i, α_i is the realized (but unobservable) value of α^*. Then in reality (9) is the expected value of y_{ij} given that $\alpha\ \alpha^*$ is α_{ij} i.e., that $\alpha^* = \alpha_i$. Hence (9) is the conditional mean.

$$E\,(y_{ij}\ \ \alpha^*{}_{=}\alpha_i) = \mu + \alpha_i \qquad \text{... (13)}$$

As already indicated, for notational simplicity we drop the use of α^* write (13) as

$$E\,(y_{ij}\ \ \alpha_i) = \mu + \alpha_i \qquad \text{... (14)}$$

Taking expectation over $\alpha_{i,}$ as in (10), then gives

$$E\,(y_{ij}) = \mu \qquad \text{... (15)}$$

Note that E $(\alpha_1) = 0$ of (10) involves no loss of generality in (14) and (15) because if E $(\alpha_1) = 0$ but E $(\alpha_1) = r$, say, then E $(y_{ij}/a1) = \mu + \tau + a1 - r$ gives E $(y_{ij}) = \mu + \tau$. Therefore, on defining μ' as μ' in place of μ and a_1' in place of α_1, and the form of (14) and (15) is retained.

Finally, we introduce the residual, similar to $e_{ij} = y_{ij} - E\ (y_{ij})$ defined earlier for fixed effects models. The definition here is

$$e_{ij} = y_{ij} - E\ (y_{ij}\ \alpha_i) = y_{ij} - (\mu + \alpha_i), \qquad \text{... (16)}$$

so that e_{ij} is a random variable; and (16) gives an equation

$$y_{ij} = \mu + \alpha_i + e_{ij}, \qquad \text{... (17)}$$

similar to (3). Then e_{ij} has properties similar to those of Example 1.

Thus

E $(e_{ij}) = 0$ and attributing uniform variance σ^2_e to each e_{ij} gives

$$\text{var}\ (e_{ij}) = \sigma^2_e = E\ (e^2_{ij}) \qquad \text{... (18)}$$

similar to (5). Furthermore, we also treat the e_{ij} s as being independent of each other and of every α_i so that

$$\text{cov}\ (e_{ij}, e_{rr}) = 0\ \forall\ i, i' \text{ and } j, j' \text{ except } i = i' \text{ and } j = j'$$

and

$$\text{cov}\ (e_{ij}, \alpha_k) = 0\ \forall\ i, j \text{ and } k. \qquad \text{... (19)}$$

Hence, for the same values of i, i', j, j' and k

$$E\ (e_{ij}, e_{rr}) = 0 \text{ and } E\ (e_{ij}\ a_k) = 0 \qquad \text{... (20)}$$

In view of (17), the variance of *yij* is

$$\text{var}\ (y_{ij}) = \text{var}\ (\mu + \alpha_1 + e_{ij}),$$

which is

$$\sigma^2_y + \sigma^2_\alpha + \sigma^2_e \quad (\text{Use (18) - (20)}) \qquad \text{... (21)}$$

In this way we see that σ^2_α and σ^2_e are components of σ^2_y, the variance of y, thus they have attracted the name *components of variance* or *variance components.*

Random Effects Models: Example 1

It is common practice for dairy farmers today, rather than mating their own bulls to their own cows, to have their cows inseminated by a technician who is supplied with bull semen from an artificial breeding corporation. That corporation's business is to own bulls that generally sire daughter cows that are high-yielding producers of milk. It can achieve that by each year buying some 80-150 young bulls that are considered to be a random sample from the population of bulls (of some particular breed-mainly Holstein, in the USA.). Then semen from those bulls is used enough so that three years later there will be approximately 60 daughter cows per bull that have milk production records. Letting yl, be the record of the jth daughter of the ith sire, an appropriate model for $\mathbf{y}_{ij}$ is then

$$y_{ij}=\mu+\alpha_i+e_{ij}$$

Heritability

To the animal breeder and farmer, who are both interested in using breeding to help increase the production of economically important products from farmed animals (e.g., bodyweight, length eggs, milk, butter, wool, tallow and bacon), the variance components $\boldsymbol{\sigma^2 e}$ and $\boldsymbol{\sigma^2\alpha}$, are of much interest. They are needed, for example, for the ratio

$$h^2 = 4\,\sigma^2\alpha\,/(\sigma^2\alpha + \sigma^2 e\,)$$

which is a parameter called heritability that is of great importance in genetics, not only in the breeding of animals but of plants too. A similar ratio is $\sigma^2\alpha$ / $(\sigma^2\alpha + \sigma^2 e)$, the intra-class correlation; it also occurs in psychological and educational testing, where it has the connotation of reliability.

Best Linear Unbiased Predictor (BLUP)

Another aspect of this model that is often of particular interest is that of predicting the value of α_i corresponding to the sire labeled **i**. This is important for ranking the sires on the basis of the predicted values of the α_i s because those predictions act as estimates of the sires' genetic merits. Given that the exact genetic contribution (a random half of his genetic make-up) of a sire to his progeny is, in fact, different for each progeny, we cannot estimate α_i in the sense that we estimate fixed effects. With α_i being a random variable, the best we can do is to consider the expected value of α_i, given the records that we have from all the progeny of all the sires. Thus we seek to estimate the conditional mean $E(\alpha_i/y)$, where $\mathbf{y}$ is the vector of all records. This estimator (which is nowadays called a predictor) turns out to be,

The estimator (21), is known as the best linear unbiased predictor (BLUP) and can also be written as

$$\tilde{\alpha}_i = \frac{n_i\sigma_\alpha^2}{\sigma_e^2 + n_i\sigma_\alpha^2}(\bar{y}_i - \tilde{\mu}) \qquad \text{... (21)}$$

The estimator (21), is known as the **best linear unbiased predictor** (BLUP) and can also be written as

$$\tilde{\alpha}1 = \frac{n_i h}{4+(n_i-1)h}(\tilde{y}_i - \tilde{\mu})$$

Estimation of Heritability

Estimation of heritability is based on methods that determine resemblance between genetically related animals. Roughly, there are two methods that can be used.

1. The resemblance between parents and offspring
2. The estimation of variance components

The Resemblance Between Parents and Offspring

If we plot the observations on offspring against the values of their parents (either sires, or dams, or their average), we can perform Offspring-Parent regression. The slope of the regression line though is plot reflects how much of the phenotypic differences that we find in parents are retrieved in their offspring. The expected value of the regression line is $b_{OP} = 0.5h^2$ (or h^2 when regression is on midparent mean). Offspring-parent regression is not often used in practice. It requires data on 2 generations, and uses only this data. It is also not able to utilize genetic relationships among parents. However, the method is robust against selection of parents.

The Estimation of Variance Components (within and between family components)

If the variation within families is large relative to differences between families, the trait must be lowly heritable. Variance components are attributed to specific effects. For example, the (paternal) half-sib variance is due to differences between sires. The variance component represents the sire variance, which is a quarter of the additive genetic variance. In analysing data, we are promptly faced with variances. With each set of data, we assume a (mixed) model that explains the observations. In this, we make a distinction between fixed effects, that determine the level (expected means) of observations, and random effects that determine variance. A model at least exists of one fixed (mean) and one random effect (residual error variance). If observations also are influenced by a genetic contribution of the animals, then a genetic variance component exists as well. In that situation, we have two components contributing to the total variance of the observations: a genetic and a residual variance component. If we calculate responses in breeding programs, we make use of those parameters. In predicting breeding values, a model can be applied which use both fixed and random effects and Best Linear Unbiased Prediction (BLUP). Variances and covariances are assumed to be known.

When to Estimate Variance Components?

In general, the estimation of variances and covariances has to be based on a sufficient amount of data. Depending on the data structure and the circumstances during measuring, estimations can be based on some hundreds (selection experiments) or more than 10,000 observations (field recorded data). It is obvious that we are not interested in estimating variance components from every data set. The information in literature is in many cases even better than estimations based on a small data set. In general, we have to estimate variance if:

- We are interested in a new trait, from which no parameters are available;
- Variances and covariances might have changed over time;
- Considerable changes have occurred in a population e.g. due to recent importations.
- Mostly it is assumed that variances and covariances, and especially the ratio of both of them (like heritability, correlation), are based on particular biological rules, which do not rapidly change over time.

However, it is well known that the genetic variance changes as consequence of selection. Changes are especially expected in situations with short generation intervals, high selection intensities or high degrees of inbreeding or in a situation in which a trait is determined by only a few genes. Secondly, the circumstances under which measurements are taken can change. If conditions are getting more uniform over time, the environmental variance decreases, and consequently the heritability increases. Thirdly, the biological interpretation of a trait can change as consequence of a changed environment; feed intake under limited feeding is not the same as feed intake under ad-lib feeding. In conclusion, there are sufficient reasons for regular estimation of (co-) variance components.

Methods Used for Estimation of Genetic Parameters - Methods of Henderson

The methods of Henderson use Least Squares equations and variance components are estimated of certain quadratics (usually differences between quadratics) and their expectations. The variance of the estimates is not minimized (i.e. the estimation is not the most accurate) because sums of squares and expectations are not dependent on the variance-covariance structure of the data but rather on LS equations. Estimates of variances are unbiased but can fall outside the parameter space (e.g. they can be negative). Estimates are also not unique because, when there are several random effects, sums of squares due to random effects can be computed in several ways, i.e. corrected for several combinations of other effects.

Methods Used for Estimation of Genetic Parameters - ML (Maximum Likelihood)

ML (Maximum Likelihood) estimators maximize the likelihood of the parameters given the density functions and the data. Estimates are not unbiased but they have smaller variance then the unbiased estimators.

Maximum Likelihood (ML) Method

Maximum Likelihood (ML) is a robust method to estimate variance components. ML estimates the parameters of distribution that maximize the likelihood of the observed data

Advantages

- Overcomes major limitations in variance estimations by ANOVA method
- ML method can jointly analyze field record with complex pedigrees
- ML estimators are ideal for unbalanced designs
- ML estimators yield variance estimates within the parameter space
- No negative variance

Drawback

- ML estimators yield biased estimates of variance components
- The residual variance tend to be downwardly biased
- The bias increases as the number of fixed effects increases
- Algebraically very tedious demanding advanced math skills in algebra, calculus and optimization

Methods Used for Estimation of Genetic Parameters - REML (Restricted ML)

REML (Restricted ML) estimators maximize the likelihood of the parameters after correcting for the fixed effects (formally: in the space orthogonal to the fixed effects). In ML methods the loss in degrees of freedom due to correction for fixed effects is not taken into account. In REML this loss in degrees of freedom is accounted for. Different quadratic forms are calculated based on the mixed model equations.

Residual Maximum Likelihood (REML) Method

REML is a restricted version of the ML method. The REML variance parameters are estimated by maximising that part which does not depend on the fixed effects (Likelihood of the random\residual effects). The restricted maximization of the likelihood eliminates the undesirable bias of the ML method. REML is the most preferred method for variance estimation

Advantage

- REML gives the unbiased estimate of variance components
- It is an improvement over the ML method

Restricted (or Residual) maximum likelihood (ReML)

ML vs. REML

The intuition behind the raw maximum likelihood method, is that it should return the estimates that makes the actual observations most likely. The intuition behind the restricted likelihood method is almost the same, but instead of optimizing the likelihood of the observations directly, it optimizes the likelihood of the full residuals. The full residuals are defined as the observations minus the fixed effects part of the model. This modification ensures, at least in balanced cases, that the random effect parameters are estimated without bias, and for this reason the REML estimator is generally preferred in mixed models.

Methods Used for Estimation of Genetic Parameters

Choice of the best method to estimated variances is not obvious. One could choose for unbiasedness but in the practice of estimating variances accuracy (minimal variance) is usually preferred. It is also important to notice that unbiased methods use least squares equations and therefore cannot correct for selection in animal breeding data, e.g. through the relationships matrix or by using correlated traits. In animal breeding, data used to estimate variance components frequently originates from selection experiments or livestock improvement schemes, which involve continuous culling of animals on the basis of their performance or breeding values. In that case, ANOVA estimators, which assume that data are randomly sampled, tend to be subject to selection bias. Under certain conditions (RE) ML will account for selection, because it makes use of the mixed model equations. This very important feature has made REML the method of choice for most animal breeding applications.

References

Cameron, N.D.1997. Selection Indices and Prediction of Genetic merit in Animal Breeding. CAB International

Estimation of Genetic Parameters. https://jvanderw.une.edu.au/Estimation_of_Variance_Components.pdf.

Neale, B.M., Ferreira, M.A.R., Medland, S.E. and Posthuma, D. 2008. Statistical Genetics. Taylor & Francis.

Searle, S. R., Charles, G. C. and Mcculloch, E. 2006. Variance Components, John Wiley & Sons, Inc., Publication.

17

Non-Parametric Data Analysis in Fisheries

Sethuraman Sivakumar P.

Parametric and Non-parametric tests

Parametric and Non-parametric tests are used for comparing differences between groups (e.g. Difference in catch-per-unit-effort (CPUE) of two types of Codends) and to test the relationship between two or more variables (e.g. Gender differences in awareness levels of fish farming (High, medium and low). As there seen differences in types of data, Non-parametric statistics are used when

- Data is non-normal
- Small sample size
- The variables are categorical
- Short response categories (<5) in scales

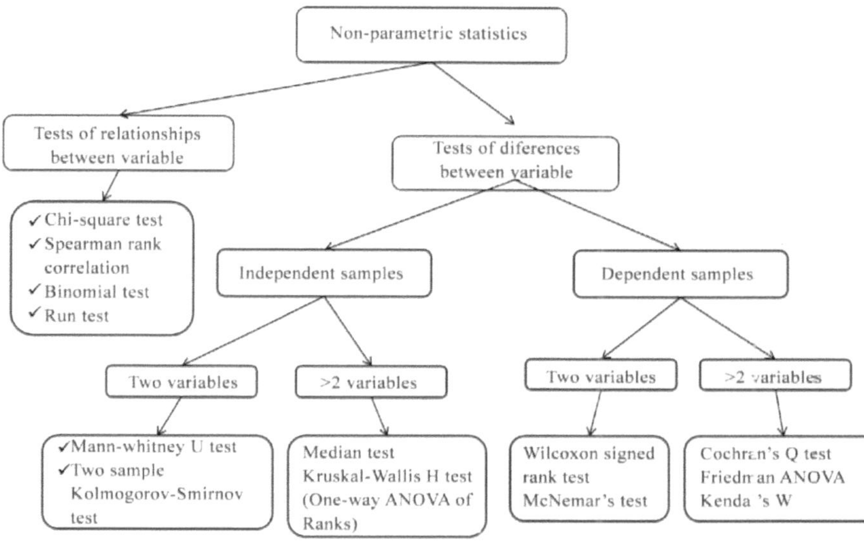

Parametric and their non-parametric equivalents

Parametric Test	Non-parametric alternative
Paired t test	Wilcoxon Rank Sum Test
Unpaired t test (Independent sample t test)	Mann-Whitney U test
Pearson Correlation	Spearman Correlation
One Way ANOVA	Kruskal Wallis Test
Two way ANOVA	Friedman test

Chi-square, Phi and Cramer's V tests

Purpose:– To determine if a statistically *significant relationship exists between the two dichotomous or nominal variables*

Sample size - large (> 50) and levels or categories (>5)

Points to Remember

- Phi and Cramer's V tests are used along with Chi-square to *estimate the strength of the association between two nominal variables.*
- If the *expected frequencies are less than 5*, the Fisher's exact test is used.

Problem

- Whether literacy level of farmers relate to their awareness on improved varieties of fish?

Results

Chi-square Tests

	Value	df	Asymp. Sig. (2-sided)	Exact Sig. (2- sided)	Exact Sig. (1- sided)
Person Chi-square	4.848[a]	1	0.028		
Continuity correction b	3.039	1	0.081		
Likelihood Ratio	5.032	1	0.025		
Fisher's Exact Test				0.065	0.040
Linear by Linear Association	4.506	1	0.032		
N of Valid Cases	20				

a. 2 cells (50.0%) have expected count less than 5. The minimum expected count is 3.60.

b. Computed only for a 2x2 table

Chi-square analysis of extension participation or non-participation among male and female farmers

Variable	N	Gender		X^2	Fishers exact significance two tailed (p)
		Male	Female		
Extension participation				4.848*	0.065
Participated	11	9	2		
Not participated	9	3	6		
Total		12	8		

* Significant at 5% level

Results displayed in the above Table indicate that the Pearson Chi-square was significant. Since the expected counts are less than 5 in two cells, Fisher's exact test was used to check the significance. The non-significant Fisher's exact test indicate that we cannot be certain that males and females are different in terms of 'participating' or 'not-participating' in extension programmes.

2. Spearman Correlation

Exists linear relationship between variables, when found Ordinal variables, non-normal data and small sample size.

Problem

Whether the superior encouragement to use computers in job and willingness for deputing employees for training are related?

Results

Correlations

			Encouragement	Deputing training
Spearman's rho	Encouragement	Correlation Coefficient Sig. (2-tailed) N	1.000** 216	0.584 0.000 216
	Deputing training	Correlation Coefficient Sig. (2-tailed) N	0.584** 0.000 216	1.000 . 216

** Correlation is significant at the 0.01 level (2-tailed)

To investigate if there was a statistically significant association between superior encouragement to use computers and their willingness for deputing employees for training, correlation was computed. Since both variables are non-normal, they violated the assumption of normality. Thus, the Spearman rho statistic was calculated, r = 0.584, p = 0.000. The direction of the correlation was positive, which means that superior encouragement to use computer is directly related to superiors' willingness to depute employees for training.

3. Wilcoxon Signed-Rank Test

Wilcoxon Signed-Rank Test is used to compare the *paired samples (Before – After)* when observations are taken on same person at two different times and it is analogous to the paired t-test.

Problem

Fish pakkoda prepared with two different oils – Mustard and Sunflower oil were tested for its taste in a five point scale (1- Very Poor to 5- Excellent)

Rationale for selection: Small sample; difference in taste scores not normally distributed.

Wilcoxon Signed-Rank Test – Normality of differences

1. Test if the differences in two samples normally distributed – SPSS Explore command (Normal QQ Plot)

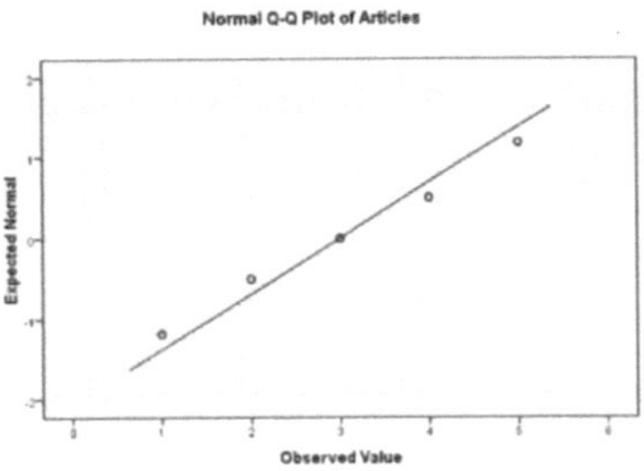

Normal Data

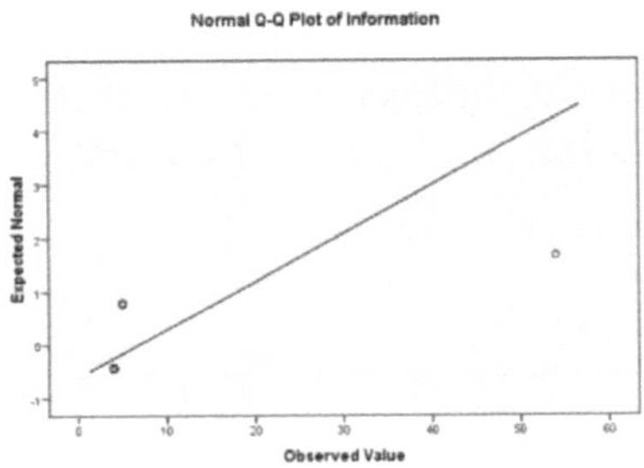

Non-normal Data

Tests of Normality

	Kolmogorov-Smirnov[a]			Shapiro-Wilk		
	Statistic	**df**	**Sig.**	**Statistic**	**df**	**Sig.**
Planning	.300	0	.000	.788	20	.001
Presentation1	.342	20	.000	.632	20	.000
Design	.288	20	.000	.746	20	.000
Information	.514	20	.000	.263	20	.000
Reports	.224	20	.010	.841	20	.004
Articles	.155	20	.200*	.896	20	.035

a. Lillefors Significance Correction

* This is a lower bound of the true significance.

Significant Value Indicates Non-normality

Results

		N	Mean Rank	Sum of Ranks
Taste rating Mustard oil- Taste rating Sunflower oil	Negative Ranks	2[a]	3.00	6.00
	Positive Ranks	g[b]	6.67	60.00
	Ties	7[c]		
	Total	18		

a. Taste rating Mustard oil < Taste rating Sunflower oil
b. Taste rating Mustard oil > Taste rating sunflower oil
c. Taste rating Mustard oil = Taste rating Sunflower oil

Test Statistics

	Taste rating Mustard oil- Taste rating Sunflower oil
Z	-2.449b
Asymp. Sig. (2-tailed)	0.14
Exact Sig. (2-tailed)	0.16
Exact Sig. (1-tailed)	.008
Point Probability	.005

a. Wilcoxon Signed Ranks Test
b. Based on negative ranks

"A Wilcoxon Signed-Ranks test indicated that the taste rating of "Sunflower oil" pakoda (mean rank = 6.67) was rated more favorably than the "Mustard oil" pakoda (mean rank = 3.0), Z = -2.24, p = 0.014.

4. Mann-Whitney- U Test

Also known as Wilcoxon rank sum test and analogous to Independent t test. The purpose of this test is to test whether two samples are likely to derive from the same population (i.e., that the two populations have the same shape) with the following assumptions.

1. The samples are independent.
2. The level of data is at least ordinal.

Problem

To test whether CPUE (kg/hr) is different for two designs of Codends

Test Statistics

	In catch-per-unit-effort
Mann-Whitney U	4.000
Wilcoxon W	14.000
Z	-1.765
Asymp. Sig. (2-tailed)	0.77
Exact Sig. [(2*(1-tailed Sig.)]	.114[b]

a. Grouping Variable: Codend

b. Not corrected for ties

Use the Exact Sig value 0.114, as Asymp. Sig is suitable only for large samples.

Result

There is no difference in CPUE of two variants of Cod ends.

5. Kendall W and Friedman ANOVA

Mean ranks and agreement coefficients of yam varietal attributes.

	Ranks			
	Subsistence		Commercial	
Varietal attribute	**Mean rank given by farmers**	**Overall rank**	**Mean rank given by farmers**	**Overall rank**
High yield	8.10[a]	2	10.35[a]	1
Early maturity	6.57[b]	5	8.65[b]	3
Dwarf plant	6.40[b]	6	4.85[c]	6
Disease resistance	7.03b	4	10.10a	2
Adaptability to soil types	8.73a	1	4.60c	7
High storability	4.27c	10	8.05b	4
Round tuber shape	5.83c	8	3.90d	9
White tuber fleshy colour	6.17b	7	4.40d	8
Smooth tuber surface	6.03b	9	2.70e	10
Free from browning	3.30c	11	2.15e	11
Culinary quality	7.57a	3	7.25b	5
Kendall W	0.25**		0.74**	
Friedman Chi-square	73.01**		148.34**	

Rank order reserved (1: least preferred; 11: most preferred)

Values followed by same letter indicates no significant difference among those attributes ($p<0.05$)

** Significant at 1% level ($p<0.01$)

Kendall W indicated the agreement among farmers in the ranking of varietal attributes-(*Range – 0 to 1)*.

Friedman ANOVA establishes the significant differences among farmers' rankings among the varietal attributes.

18

Cutting Edge Technologies in Fisheries

Murthy G.R.K.

Introduction

Resources are used unscientifically commonly. For example applying feed somewhere in the pond results in loss, but by using technology, we can know when to apply where to apply, and how to apply which leads to efficient management of resources. Monitoring should be continuous which is not possible and very difficult for humans, but with technology, it could be monitored and the status could be known even from home which is communicated through mobile, which is sent by sensors installed. The data are also easily maintained. When technology is used, resources are efficiently utilized, since everything has a demand and that can be satisfied. So resources can easily be maintained which reduces wastage. Field management is a very skilled activity and it needs years of experience and onsite exposure which needs skilled labour, in such cases technologies aids with better methodologies. When the resources are applied at the right time, at the right amount, the growth becomes normal which increases the produce quality. After a product is produced, it has to be sent to a series of channels such as transport, market etc. and a technology should backup and monitor everything. The produce after being produced should be sold and many apps can be downloaded to improve the marketing. The whole world is changing within a period of time. So one should stay updated to impress the livelihood and the wide knowledge on technology helps to lead a better life with easy management practices. Thus data driven technology interventions are needed for efficient management of resources, monitoring of things anytime & anywhere, reduction of wastage of resources, reduction of dependence on trained and skilled manpower, enhanced product quality, better integration, better marketability, to stay updated with latest knowledge and access to other knowledge source for better upgrades.

Few Real Time Fisheries Examples in Indian Context

a) ***Yeruvaka:*** It is a Cloud-based Pond Management Solutions for aquaculture farmers, to reduce their risk and increase productivity by integrating sensors, mobile connectivity and decision tools like shrimp auto-feeder with intelligent feeding sensor. Solar powered, real-time monitoring of DO, automatic feeder for shrimps, can be remotely controlled, up to 4 days of battery backup, 125 kg bin capacity, spreads feed up to 12m radius uniformly through which the yield and profit could be predicted.

b) ***Farmmojo:*** It's an AI powered shrimp farming management app-Ai advisor records production data, water quality parameters, feed inputs, health status and biomass conversion. Collected data is analysed on a real-time basis and advisory alerts are offered to farmers about their pond operations.

Modern technology interventions for Agriculture and allied sectors

a) ***Internet of Things (IoT):*** It's a system of interrelated computing devices, mechanical and digital machines, objects, animals or people that are provided with unique identifiers. I has the ability to transfer data over network without human to human or human to computer interaction. The thing in IoT can be a person with heart monitor implant, a farm animal with a biochip transponder, an automobile with built-in sensors to alert the driver when the tide pressure is low, any other natural or man-made object that can be assigned as Internal Protocol (IP) address and is able to transfer data over a network.

IoT in Fisheries

Assessment of water quality in terms of dissolved oxygen level, pH level, alkalinity and temperature can be measured through sensors. The sensors can be connected to processing devices such as Arduino board that sends data to central database through network protocols and also can send SMS alerts to the user by GSM modules. Sensor equipped drones and robots can also collect data such as water pH, salinity, DO levels turbidity, pollutants and even the heart rate of the stock and all these could be accessed via a smartphone.

b) ***Artificial Intelligence (AI):*** AI is human-like that can simulate human intellect and behaviour. It can solve mathematical problems like puzzles games theorems, common sense reasoning, expert knowledge i.e., knowledge of a lawyer, medicine, diagnosis, agricultural practices etc. It is innovative and exhibits social behaviour and it can achieve the set goals and have performance measure.

Machine Learning Vs Deep Learning

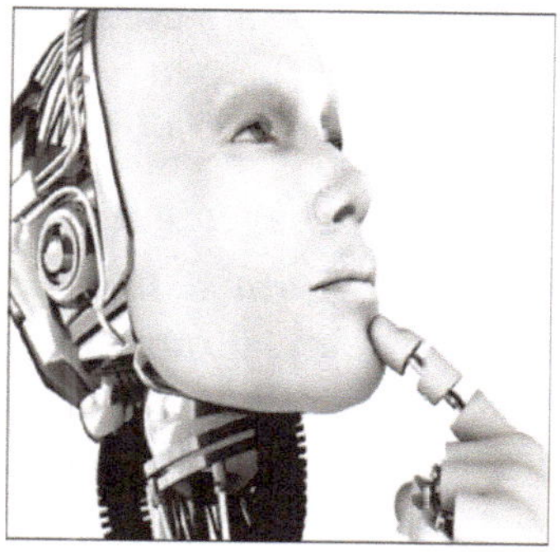

Machine Learning learns from the data and then apply what they have learned to make an informed decision. It teaches machines instead of programming them. Big data can train machine learning models. Deep Learning structures algorithms in layers to create an "artificial neural network" that can learn and make intelligent decisions on its own. Machine Learning and Deep Learning are ways of achieving AI.

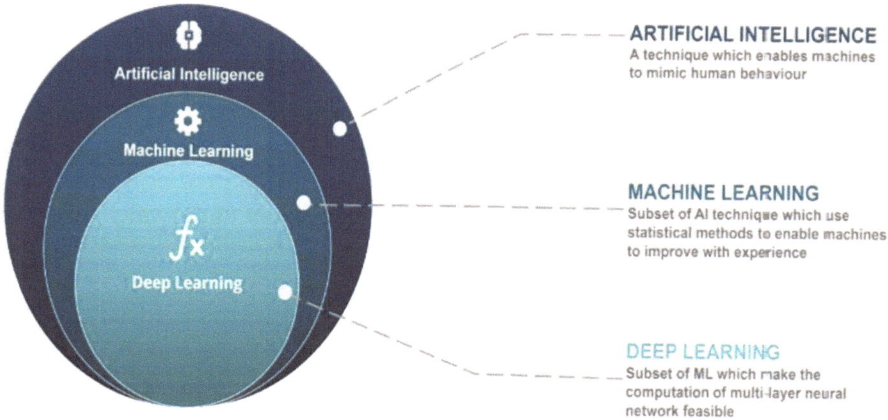

AI and IoT Application in Fisheries

AI and IoT finds application in various sectors of fisheries, *viz.,* feed management, disease management etc. which are detailed below.

Feed Management

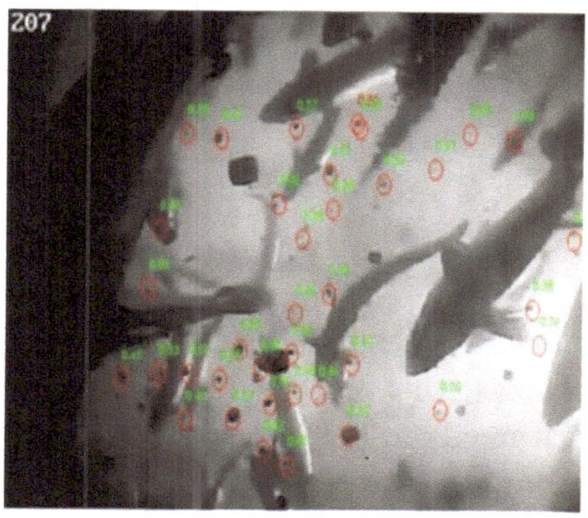

It is clearly evident that, skill of gauging on feed management takes years to develop, but AI and data processing systems can track measurable patterns when the stocks are feeding. AI of constantly learning process gauges why feed pellets remain uneaten for a certain point that led to manually or automatically adjust the amount of feed resulting in waste minimization and profit maximization. In a Fishery system, sensors are used to detect the hunger levels in shrimp and fish, and the controlling dispensers release the right amounts of food at the right time and reduces the feed cost by up to 21%.

Disease Management

Norway's Seafood Innovation Cluster: Worked with IBM to build Aqua-Cloud, a new platform that gathers data from salmon farms across Norway and uses sophisticated machine learning techniques to help predict and prevent sea lice outbreaks. AquaCloud is a cloud based program that aims to help fish health managers and researchers deal with sea lice, predicting or even preventing lice development in sea cages with the goal of reducing dependency on expensive medical treatments, thus minimising stock mortality.

Aquaconnect: India's trusted aquaculture platform provides end-to-end solutions to shrimp and fish farmers powered by AI and satellite remote sensing technologies.

FarmMOJO: An AI-powered mobile app to improve efficiency in aqua-farming. The mobile application help shrimp farmers to predict diseases and enhance water quality. SHOAL uses robo-fish to detect underwater pollution sources near fish farms and other facilities.

XpertSea: Uses computer vision and AI to calculate the growth of shrimp, helps farmers to predict the most profitable harvest periods. XperCount collects critical animal data by using cameras and machine learning which is applied to count, size and weigh animals in seconds. By combining this shrimp growth data with market prices, making it much easier for farmers to make the right decisions. Farmers can remotely switch pumps, motors, aerators or diffusers on or off. Production and demand can be forecasted by altering program parameters, further improving farm efficiency and monitoring ability. Advanced AI techniques like deep learning are used to pinpoint timeframes by continuously using machine learning on historical growth cycle data. Farmers are able to track the growth of their crops, feeding regimes and up to 20 water parameters and predict shrimp growth 14 days in advance, based on existing data.

Block Chain Technology

A Block chain is a growing list of records called blocks, which are linked using cryptography. Each block contains a cryptographic hash of the previous block, a timestamp, and transaction data. Decentralized, distributed, encrypted database that records data, or it is a digital ledger of any transactions, contracts that needs to be independently recorded. Some of the examples are listed as follows:

- Food Safety – What is the source of my food?
- Food Supply Chain – Who is making the most profit?
- Agri Inputs – Are the inputs authentic?
- Land Title Registering – Is the land I am buying or leasing litigation free?
- Disbursement of Subsidies – Is the benefit really reaching the farmer?
- Opening New Markets for Farmers – Can farmers sell directly to us online?
- Community Agriculture
- Accountable Relief Distribution
- Block Chain technology

Characteristics of Block Chain

Tamper-proof data	Shared & Transparent	Consensus Based	Security & Privacy
• Records written to the Block chain cannot easily be changed or deleted • Enhances trust in decentralized single source	• Distributed ledger enables single shared source of truth distributed across all nodes • Enables transparency, reduced delays & cost of 3rd party intermediaries	• Updates to distributed ledger are made after consensus among network peers • Protects against fraud among participants, and maintains trust in the ledger	• Authorized with keys only can view the data • Block chain stores auditable historic records that can be encrypted for privacy

Why Block Chain Technology is Used in Seafood Sector?

In seafood sector, BCT is used in tracking the traceability of the origin of materials and parts; the product processing history; and the distribution and location of the product after delivery. In traceability, there is no guarantee that the products are true or complete, as both error and fraud can lead to false claims about the properties of the food product, including its origin. Traceability needs to deal with issues such as lack of transparency, fraud and species substitution, integrity and seafood safety. Ensuring traceability through the seafood production chain can be accomplished by careful planning, taking the time to gain consensus among the operators and authorities.

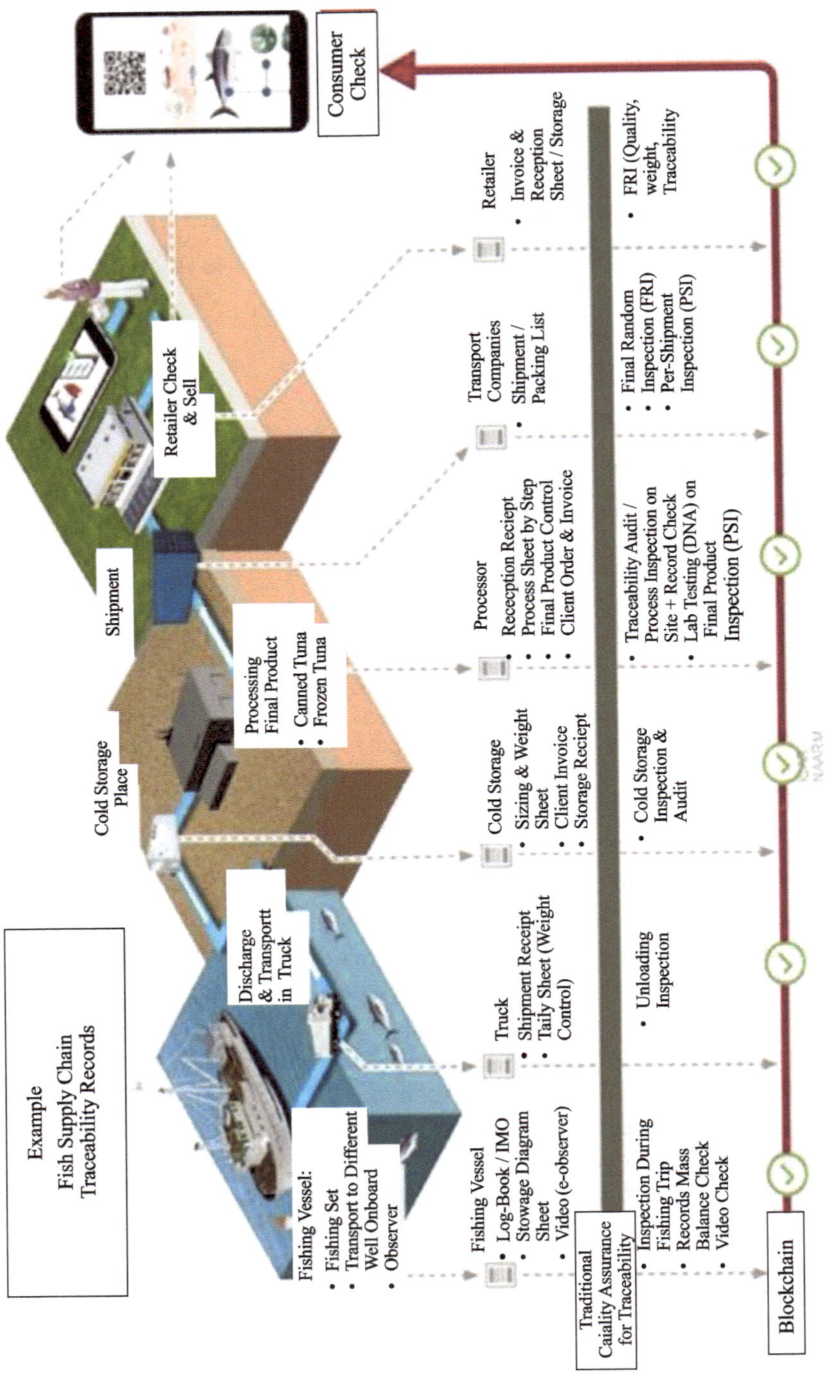

Development and Operational Considerations in Block Chain Technology

Present Status of Block Chain in the Wild-caught Seafood Value Chain

Provenance project: Earliest Block chain project in seafood value chains was in 2016 by Project Provenance Ltd (Provenance) in Indonesia. It focused on supply chains of yellow fin tuna loins and skipjack tuna for canning. On capture, catch is registered by SMS message and physically tagged to identify new asset on the block chain. Using Provenance application on-land catch is transferred to suppliers both physically and digitally. Provenance integrated with at least one ERP system, TallyO helps to track the catch through the processing stages and out to shipment. Using NFC-enabled smart labels, product packaging with smart labels helps to communicate provenance story.

WWF-New Zealand, ConsenSys, Sea Quest (Fiji) Ltd., and Tra Seable Solutions

This is to create a completely transparent and traceable supply chain by utilising private block chain technology for the fresh and frozen tuna supply chain. Supply chain is mapped into Treum platform, and the roles and permissions will be set accordingly. On capture, each tuna is tagged with unique identifiers initially using RFID tags and the data are transmitted in real time to the block chain using internet. On landing, each unloaded tuna is tracked by scanning its tag. If a tuna is transformed into other products such as loins, then each new product (loin) will be given a new identity on the block chain and tracked separately. It uses Quick Response (QR) codes on product packaging to communicate provenance story.

Pacifical and Atato

First large-scale public block chain initiative to track purse seine tuna destined for canning in Thailand during 2018. In this, fish are not individually identified. The platform used is Atato notary application programming interfaces. Uses existing fisheries information management system platform for data capture of Marine Stewardship Council. Atato notary service receives digital traceability data at key points and records onto block chain. Provenance story linked to lot/batch number printed on canned tuna.

OpenSC – WWF-Australia and BCG Digital Ventures

For traceability of Patagonian tooth fish in 2019. Integration of machine learning with GPS data to determine whether the fish has been caught in a legal area. Tagging of tooth fish with RFID tags on capture and recorded data will give the information about the movement of fish through the supply chain.

Bumble Bee Foods

Bumble Bee Foods is a private block chain network that uses SAP's Blockchain services to track "Fair Trade" certified yellow fin tuna sourced from Indonesia.

Uses QR codes on the tuna product packaging to communicate the provenance story of the fish.

Fishcoin

Fishcoin is a block chain based data ecosystem (2018) for seafood products. It's a series of open source tools and software development kits. Actors exchange Fishcoin tokens with the previous actor for the catch data, until it reaches the retailer that sells it to the consumer. Every actor in custody of the fish adds more data to the ecosystem.

Sustainable Shrimp Partnership

Sustainable Shrimp Partnership is a consortium based block chain network. IBM Food Trust block chain platform provides transparency and traceability for its Ecuadorian farmed shrimp, because the producer data about products are accessible to retailers around the world.

Challenges across the Projects

a) **Reliant on human input:** Most of the projects rely on human input of fish data, which themselves could be open to tampering

b) **Tagging and labelling of fish:** Physical fish tags/labels could be lost or damaged while transporting the fish or could potentially be tampered with.

c) **Verifiability of private and consortium block chain platforms:** These types of block chains are not open to the public and transactions on them cannot be independently verified.

d) **Complex seafood value chain scenarios untested:** The solutions are not tested in real-world complex seafood value chain scenarios where the value chain actors are unknown.

References

http://www.fao.org/3/ca8751en/ca8751en.pdf (Blockchain Application in Sea Food Value Chains)

https://thefishsite.com/articles/a-practical-guide-to-using-ai-in-aquaculture

19

Methodologies for Assessing Climate Change Impacts on Agriculture, Fisheries and Aquaculture

Saravanakumar V.

Key Facts

- Global fish production peaked at about 171 million tonnes (MT) in 2016, with aquaculture representing 47 perc ent of the total and 53 per cent of non-food uses (Global capture fisheries production was 90.9 MT in 2016)
- World fish production is projected to increase by 15% in the next 10 years, reaching around 200 MT per year
- The total first sale value of fisheries and aquaculture production in 2016 was estimated at USD 362 billion, of which USD 232 billion was from aquaculture production
- Fish (including shellfish) provides essential nutrition for 3 billion people and at least 50% of animal protein and minerals to 400 million people from the poorest countries
- 10-12% people, i.e. over 870 million people, depend on fisheries and aquaculture
- Aquaculture is the world's fastest growing food production system, growing at 7% annually
- Women account for 19% of all people directly engaged in the fisheries and aquaculture sector, and over 50% when including the post-harvest sector

Climate Change: A global externality

Climate Change (CC) is a result of the externality associated with greenhouse-gas emissions. It entails costs that are not paid for by those who create the emissions.

Distinguished Features from Other Externalities

- Climate change is global in its causes and consequences;
- The impacts of climate change are long-term and persistent;
- Uncertainties and risks in the economic impact are pervasive;
- There is a serious risk of major, irreversible change with non- marginal economic effects

Definition of Climate Change

Climate change refers to a statistically significant variation in either the mean state of the climate or in its variability, persisting for an extended period (typically decades or longer).

Climate change may be due to natural internal processes or external forcings, or to persistent anthropogenic changes in the composition of the atmosphere or in land use (IPCC).

Framework Convention on Climate Change (UNFCCC), in its Article 1, defines "climate change" as: "a change of climate which is attributed directly or indirectly to human activity that alters the composition of the global atmosphere and which is in addition to natural climate variability observed over comparable time periods".

Difference between Climate variability and Climate Change

Climate variability	**Climate Change**
The way climatic variables (such as temperature and precipitation) depart from some average state, either above or below the average value	Climate change can be defined as a trend in one or more climatic variables characterized by a fairly smooth continuous increase or decrease of the average value during the period of record.
Although daily weather data depart from the climatic mean, we consider the climate to be stable if the long-term average does not significantly change	Long-term average does significantly change
As we look at 30-year average values, however, we detect climate variability	For example, the 30-year average July temperature from 1971 to 2000 is lower by approximately 1°F than that of 1941 to 1970.
Short-term phenomenon	Long-term phenomenon

Climate change may result from

- Natural factors, such as changes in the Sun's energy or slow changes in the Earth's orbit around the Sun;
- Natural processes within the climate system (e.g., changes in ocean circulation);

- Human activities that change the atmosphere's makeup (e.g. burning fossil fuels) and the land surface (e.g. cutting down forests, planting trees, building developments in cities and suburbs, etc.).

GHG Accumulation and Global Warming Changes in Fisheries

The following are the major factors / variables that have to be included in CC impact assessment in fisheries.

1. Glaciers
2. Oceans currents
3. El Niño Southern Oscillation
4. Sea level rise
5. Rainfall
6. River flows
7. Lake levels
8. Thermal structure
9. Storm severity
10. Storm frequency
11. Acidification

The impact of climate change on Fisheries and Aquaculture is depicted in following figures:

Marine Ecosystems

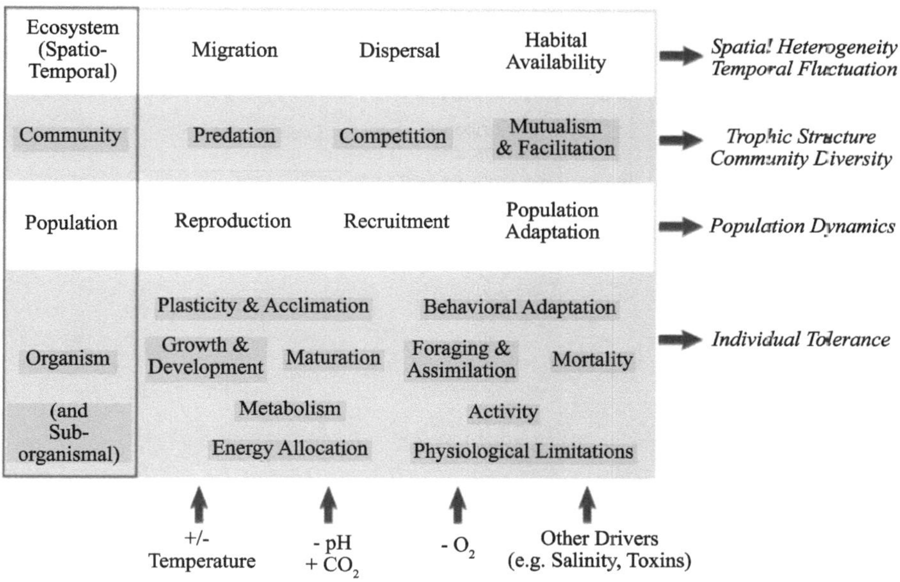

Fisheries and Aquaculture

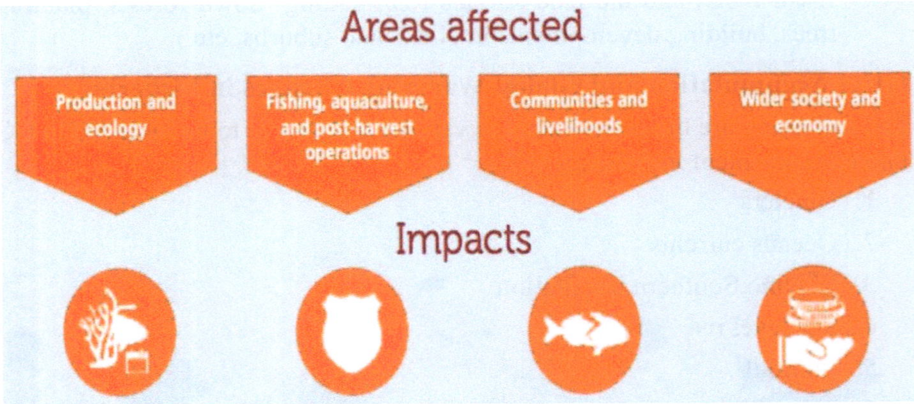

Species Composition Production and Yield, Distribution and Seasonality, Disease and Other Disruptions, Coral Bleaching, Calcification.

Safety end Security, Efficiency and Costs, Infrastructure Security.

Loss and Damages to Assets, Risks to Life and Health, Vulnerability and Confidence, Displacement and Conflict.

Cost of Migration and Adaptation, Social and Market Impacts, Water and Other Resources.

Economic impact assessment approaches

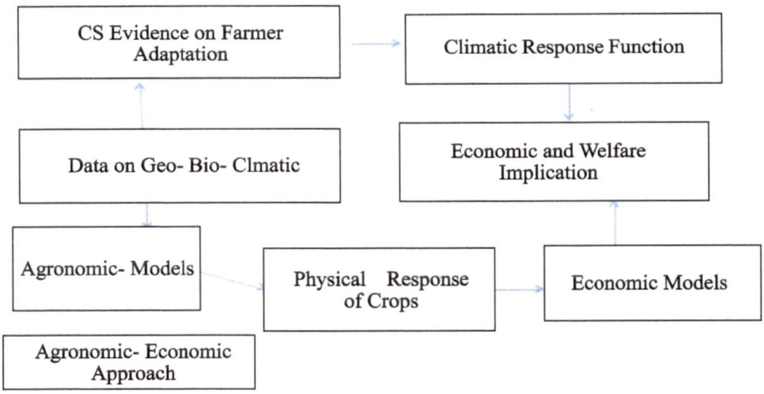

Types of relationship between rising damages and sectoral impacts of climate change

Type of effect	Sector	Functional form	Basis
Climate system	Water	Exponential $y=e^x$	The clausious-clapeyron equation shows that the water holding capacity of air increases exponentially with temperature. This means that the water cycle will intensify leading to more severe floods and droughts. There will also be more energy to drive storms and hurricanes.
	Extreme temperatures (threshold effects)	Convex curve (i.e. gradient increases with temperature)	Because of the shape of the normal distribution, a small increase in the mean dramatically increases the frequency of an extreme event.
Physical impacts	Agricultural production	Inverse parabolic (hill function) $y=-x^2$	In cooler regions, low levels of warming may improve conditions for crop growth (extended growing season and new areas opened up for production), but further warming will have increasingly negative impacts as critical temperature thresholds are crossed more often. Tropical regions may already be past the peak. The shape and location of the curve depend on crop.
	Heat-related human mortality	U-shaped	Sharp increase in mortality once human temperature tolerance is exceeded (heat waves and cold-snaps). Initially mortality will be reduced by warming in cold regions.
	Storm damage	Cubic $y=x^3$	Infrastructure damage increases as a cube of wind-speed
Human Response	Costs of coastal protection	Parabolic $y=x^2$	Costs of sea-wall construction increase as a square of defense height

Data on Geo-bio-climate will act as input for the agronomic models and through these models physical response of crops could be used and in-turn for economic models. From the economic models, economic and welfare implication could be assessed. Through the cross-sectional data, the adaptation strategies could also be assessed leading to climate response function and end up in economic and welfare implications.

Methodologies of CC Impacts

The following are some of the approaches used in Agriculture and Aquaculture for CC impact assessment; Particularly from Economic models to Dynamic size-based food web model, all have been used for the CC impact assessment in Fisheries and Aquaculture.

- Crop simulation models
- Agro-Ecological Zone (AEZ) approach
- Economic Methods
- The Earth System Model
- Dynamic Bio-climate Envelope Models (DBEM)
- Dynamic size-based food web model

Crop Simulation Models

Crop simulation models draw on controlled experiments where crops are grown in field or laboratory settings simulating different climates and levels of CO_2 in order to estimate yield responses of a specific crop variety to certain climates, and other variables of interest. These models do not include farmer adaptation to changing climate conditions in the estimates. Consequently, their results tend to overstate the damages of climate change to agricultural production (Mendelsohn and Dinar 1999). For *e.g.* InfoCrop and the main features of it are:

- Crop growth
- Water deficit
- Nitrogen management
- Temperature
- Crop-pest interactions, soil water and nitrogen balance
- (soil) OC dynamics
- GHG
- Aerobic and anaerobic conditions

Agro Ecological Zone (AEZ) Approach

Combines crop simulation models with land management decision analysis, and captures the changes in agro-climatic resources (Darwin *et al.*, 1995; Fishcher *et al.*, 2005). AEZ analysis categorizes existing lands by agro-ecological zones, which differ in the length of growing period and climatic zone.

The length of growing period is defined based on temperature, precipitation, soil characteristics, and topography. Limitations of AEZ are i) Simulate a laboratory-type set up and provide data on climatic factors and crop growth and ii) Do not take adaptive behavior of an optimizing farmer into account.

Economic Methods

a) Correlation and Regression

Data on yield and different climatic variables can be plotted to see the relationship between the said items and also about the casual effect.

Correlation Betweem Climate and Production in Tamil Nadu from 1971-2009						
Crops	**Rice**	**Sorghum**	**Spiked Millet**	**Cereals**	**Pulses**	**Food Grains**
Rainfall Deviation	0.09	-0.01	-0.02	0.18	-0.10	0.09
Total Rainfall	0.11	-0.06	-0.03	0.16	-0.14	0.07
Maximum Temp.	-0.28	-0.24	-0.40	-0.23	0.08	-0.04
Minimum Temp.	-0.32	-0.64	-0.70	-0.38	-0.18	-0.19

Total RF Vs Rice Production

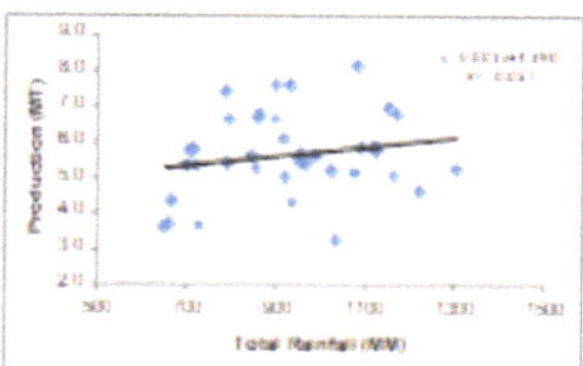

% Deviation of Deficit RF Vs Rice Production

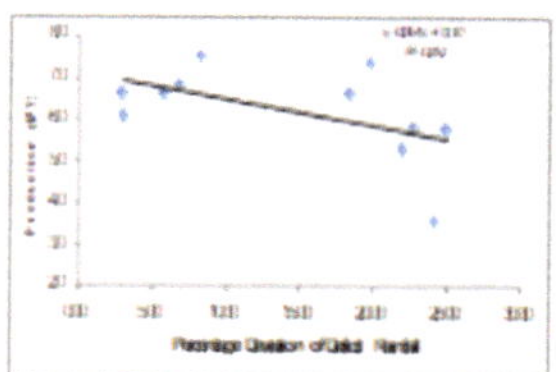

% Deviation of Excess RF Vs Rice Production

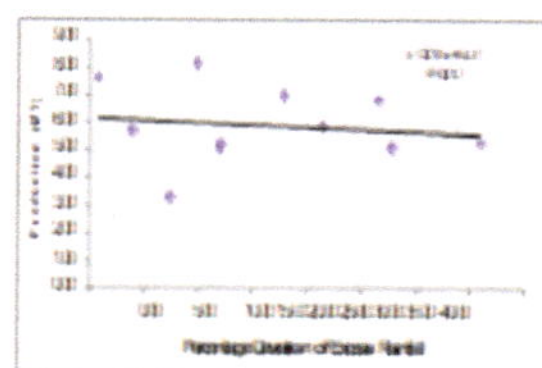

b) Production Function Approach (PFA)

Production function approach explains the relationship between production and climate which could be calibrated through experiments and used to predict changes in yield (Adams *et al.,* 1989; Rosenzweig and Parry, 1993). The outcome of the simulations incorporated into an economic model of farmer behavior, which in turn leads to a partial equilibrium model of farm sector. It is structural model which predicts two basic phenomenon: Yields (Q_i) and crop shares (H_i). Rosenzweig and Iglesias, 1994, used yield per hectare of each crop likely to be a function of climate, soil and other factors.

$Q_i = Q_i(K_i, E)$ i=1,…n

where,

K_i is a vector of all purchased inputs

E is a vector of exogenous environmental inputs

$Q_i = \hat{o}_i(K_i, E) + u_{1i}$, $\hat{o}_i$ -Predicted value of Q_i Crop shares function

$H_i = H_i(E,Z,Q_i)+u_{2i}$

Fraction of hectares of each crop (Hi) in an area is a function of Climate (E), Market access (Z) and yield (Q)

Profit maximizing criteria Max $P= S\ H_i(P_i Q_i - S\ w_j k_j)$

Limitations of PFA

Production function approach is close to collaboration with agronomic science. However, although this approach isolates the impact of environmental change, it does not fully account for the myriad of adaptations that farmers may make in response to varying environment conditions.

Ricardian Approach

Ricardian approach, instead of looking at the yields of specific crops, examines how climate in different places affects the net rent or value of farm land (Mendelsohn *et al.,* 1994). This approach takes into account both the direct impacts of climate on yields of different crops as well as the indirect substitution to other activities, introduction of new land uses, and other potential adaptations to different climates.

Crop Switching

Adaptations are voluntary, farmers will choose to undertake them only if they are beneficial. Thus to fully take into account, both the costs and benefits of adaptation are considered, so the relevant dependent variable should be NET REVENUE.

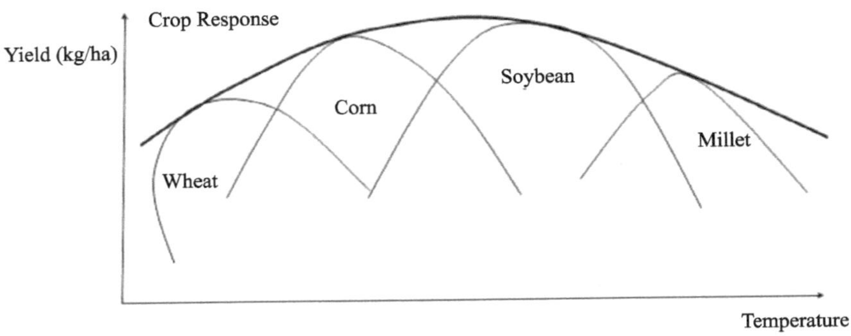

Ricardian Model

Production function $Q_i = Q_i(K_i, E) \quad i = 1, \ldots, n$

Cost function $C_i = C_i(Q_i, w, E)$

Inverse Demand function $P_i = D^{-1}(Q_1, \ldots Q_i, \ldots Q_n, Y)$

Max $Q_{i,}$ Profit = $P_i Q_i$ - $C_{i-}(Q_{i,}$ w, E) - $P_L L_i$ ($P_{L\ \text{annual cost of rent}}$)

Perfect competition in Land market drive profits to zero

$P_i Q_i^*$ - $C_{i-}^*(Q_i^*$, w, E) - $P_L L_i^*$ = 0

Land rent per hectare or Net Revenue per hectare

$P_{L=}$ [$P_i Q_i^*$ - $C_{i-}^*(Q_i^*$, w, E)] / L_i^*

where,

Q_i – Production

$P_i Q_i$ – Total Returns (Price per unit * Output)

C_i – Cost of input

K_i – Purchased input

w – Input cost

E – Environmental input

Using the duality, cost function could be derived and from the original demand function, inverse demand function could be written. From the inverse demand function, profit maximization could be estimated.

Present value of current stream

$$V_L = \int_0^{\infty} P_L e^{-rt}\, dt = \int_0^{\infty} \Sigma Q_i - C_{i-*}\left(Q_{i*}, w, E\right) e_{-rt} / L_i dt$$

Change in Annual welfare from net environmental damage.

$$\Delta W = W(E_B) - W(E_A) =$$

$$\int_0^{\infty} \Sigma D^{-1}(Q_i) dQ_i - \Sigma C_i - (Q_i, w, E_B) - \int_0^{\infty} \Sigma D^{-1}(Q_i) dQ_i - \Sigma\, C_i - (Q_i, w, E_A)$$

$$\otimes W = W(E_B) - W(E_A) = P_{LB} L_B - P_{LA} L_A$$

The present value of the welfare change

$$\int_0^{\infty} \Delta W e^{-rt}\, dt = V_{LB} L_B - V_{LA} L_A$$

Empirical Model

$V_{net} = \beta_0 + \beta_1 Td + \beta_2 Td^2 + \beta_3 Tr + \beta_4 Tr^2 + \beta_5 Rd + \beta_6 Rd^2 + \beta_7 Rr + \beta_8 R_r^{2} + \beta_9 TdRd + \beta_{10} TrRr + \Sigma cjGj + \mu$

where,

Vnet is farmland net revenue,

T and R are the mean temperature and the mean rainfall, respectively,

r represents rainy season and d dry season,

Gj is a set of socio-economic characteristics of the household,

bi, ai and cj are coeffcients of the variables, β_0 is a constant term, and
μ is an error term

Limitations

- Availability of land prices as well as non- existence of efficient land markets are two major obstacles in applying the Ricardian method to most of the developing countries.
- Existence of omitted variable problem, because it does not take time-independent location-specific factors such as unobservable skills of farmers and soil quality into account
- Additionally, yield variability has been found significant in many other studies, but a Ricardian model is not capable of capturing the effect of changes in climatic factors on it (Mearns *et al.,* 1997).

Naive vs Clairvoyant Farmer

Two extremes (naive farmer (or, analyst?) Vs. clairvoyant farmer), lead to systematically lower biased estimates in Ricardian approach. Ricardian approach is criticized to be based on so-called homo economicus assumptions of utility maximizing and perfectly rational economic agents i.e., farmers are assumed to identify instantaneously and perfectly any change in climate, evaluate all associated changes and then modify their actions to maximize profits. These assumptions also imply that agricultural system is ergodic i.e., space and time are substitutable. Ergodic assumption imply, for example, that skills, institutional and financial endowments for responding to drought (that are typically refined in arid places) are assumed to be available for use by people in humid areas (where such resources are under-developed) immediately and in essentially cost-less manner. Thus Ricardian climate sensitivity estimated for one point in time or place may not represent sensitivities for other times or places.

Ricardian Approach with Space

Spatial lag operators can be introduced in the analysis to capture information diffusion process and inter-farmer communication. A significant coefficient on spatial lag operator indicates that land use in a region is influenced by land uses in adjacent regions. Similarly group-wise heteroscedastic terms can be introduced to account for differences between broad regions (meso-scale) (Schlenker *et al.,* 2005, 2006; Polsky, 2004). Analysis in Indian context indicated significant influence of inter-farmer communication and inter-regional differences on climate sensitivities.

Climate Sensitivity and Spatial Dependence

Spatial-error model

$y = X\beta + \eta,\ where\ \eta = \rho W\eta + \varepsilon$

Spatial-lag model

$y = \rho Wy + X\beta + \varepsilon$

where,

y is (nx1) the vector of dependent variable observations, X is (nxm) the matrix of observations on independent variables including the climate and other control variables,

β is (mx1) the regression coefficient vector, η is (nx1) the vector of spatially correlated error terms,

ρ is (1x1) the spatial autoregressive parameter, *W is (nxn) the spatial weights matrix, and*

e is (nx1) the vector of random error terms.

Spatial Dependence

High-high spatial dependence is found among parts of western Tamil Nadu and Karnataka, northern Uttar Pradesh and Punjab; whereas low-low spatial dependence is found among south central and western (Gujarat) parts of India.

Ricardian Base Model

R= f (T_j, T^2, P ,P^2, T , P , DT, YT, YP, SOIL, BULLOCK, TRACTOR

POPDEN, LITPROP, CULTIV, HYV, IRR, LAT, ALT)

Spatial Lag Model

R= f (T_j, T^2, P ,P^2, T , P , DT, YT, YP, SOIL, BULLOCK, TRACTOR

POPDEN, LITPROP, CULTIV, HYV, IRR, LAT, ALT, RLAG)

Spatial Error Model

R= f(T_j, T^2, P ,P^2, T , P , DT, YT, YP, SOIL, BULLOCK, TRACTOR,

POPDEN, LITPROP, CULTIV, HYV, IRR, LAT, ALT, ULAG)

DUMP Farmer Approach

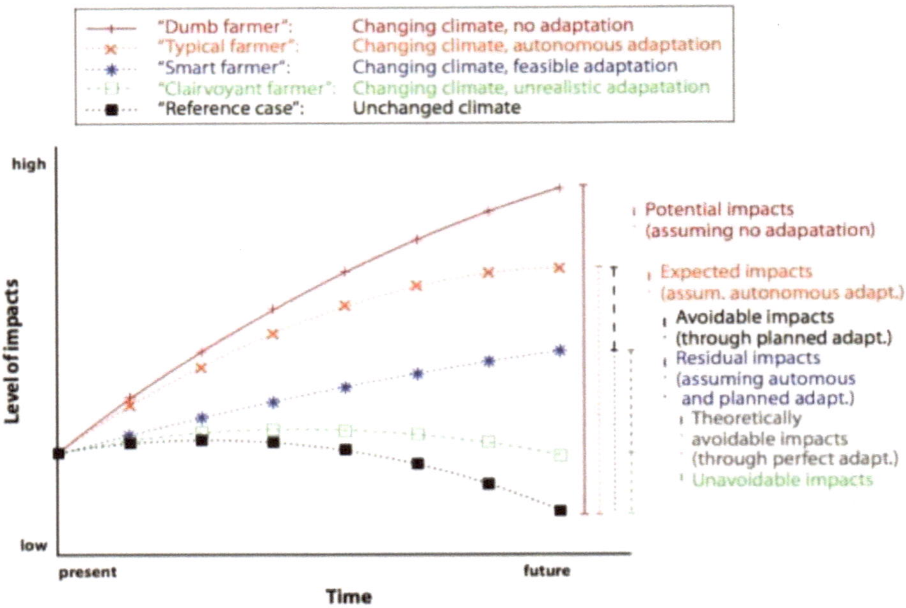

Just-Pope Production Function

$Y_{it} = f(x_{it};\beta)+\omega_{it}h(x_{it};\delta)^{0.5}$

where,

- ω_{it} is the stochastic term with mean zero & variance $_{\sigma}^{2}$
- B and δ are the production function parameters
- The independent variables (x_{it}) includes a constant, annual precipitation(P), temperature(T), trend(t) & regional dummy variables
- The expected crop productivity is given by $E(y_{it})=f(x_{it};\beta)$
- The crop variability is given by $V(y_{it})=\sigma^2 h(x\ ;\delta)\ \omega$ it

The functions $f(x_{it};\beta)$ & $h(x_{it};\delta)$ are the mean and variance functions, respectively. The derivatives of the variance function $h(x_{it};\delta)$ *wrt* the input variables *viz.,* precipitation and temperature can be used to identify whether a climate variable increases or decreases crop variability. So, if $h_x > 0$, it indicates the input variable is risk increasing, if $h_x < 0$, it implies risk decreasing. Just-Pope Stochastic Production Function helps to estimate yield variability and effect of an input variable on risk. The advantage of this PF is that the riskiness of an input variable can be readily obtained by the sign of the coefficient of that variable in the function.

Just-Pope Method

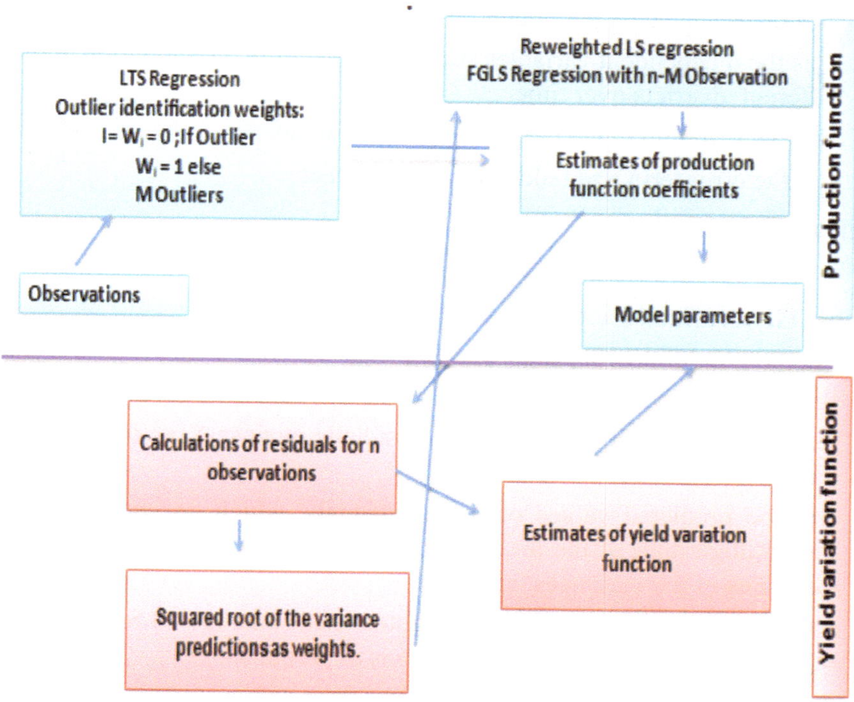

Different Functional Forms

- Linear form

$$f(x_{it};\beta,d)=\beta_0+\sum_{i=1}^{n}\alpha_i X_i+\beta_1 P+\beta_2 T+\beta_3 t+\sum_{i=1}^{i=R-1} d_i D_i$$

- Quadratic form

$$f(x_{it};\beta,d)=\beta_0+\sum_{i=1}^{n}\alpha iXi+\beta_1 P+\beta_2 T+\beta_3 t+\beta P^2+\beta P_6 T+\sum_{i=1}^{i=R-1} dD_{i\ \ i}$$

- Variance Function

$$h(x_{it};\delta,n)=\exp(\delta x_{it}+nD)=\exp\delta+\delta_1 P+\delta_2 T+\delta_3 T+\sum_{i=1}^{i=R-1} n_i D_i \ni 0'_f$$

Panel Data Model

$y_{it}= c_i + \theta_i + {}_i t + Z_{it}\beta + {}_{it}$

where,

- y_{it} is the district yield of major food crops overtime

- c_i and θ_i are fixed effects for districts over different years
- $_i$ is a district specific parameter on the annual time trend t
- Z_{it} is the independent variables such as temperature, precipitation across different districts over time
- Vectors β are unknown parameters to be estimated
- ε_{it} is the error term and boldface indicates vector

Quadratic Model

$$y_{it} = c_i + \theta_i + \varphi_i t + T_{it}\beta_1 = T^2{}_{it}\beta_2 + P_{it}\beta_3 + p2_{it}\beta_4 + T_{it} * P_{it}\beta_5 + \varepsilon_{it}$$

Feasible Generalized Least Square (FGLS) Method

In this method, the following steps are followed.

- Estimate the model by ordinary least squares (OLS) and get the residuals;
- Regress the logarithm of squared residuals against X as independent variables;
- Get the predicted values of those residuals, which are calculated as the antilogarithm of the predictions from step (b). They are consistent estimates of the variances; and
- Estimate the original model by weighted least squares (WLS), using the squared root of the variance predictions as weights.

Panel Corrected Standard Errors

Beck and Katz (1995) (henceforth, BK) use Monte Carlo methods to study the performance of FGLS in a statistical environment characterized by (i) group-wise heteroscedasticity, (ii) first-order serial correlation, and (iii) contemporaneous cross-sectional correlation. FGLS (Parks) produces dramatically inaccurate coefficient standard errors.

Maximum Likelihood Estimation

The likelihood function is given by

$$nL = -1/2\overset{\gamma}{\underset{\leq}{,}}\ NIn(2\pi) + \sum_{t=1}^{T}\sum_{i=1}^{R} In(h(x_{it} - \delta)) + \sum_{t=1}^{T}\sum_{i=1}^{R}\{yit - f(x_{it} - \beta)\}^2 h(x_{it};\delta)_{\infty}^{/}$$

where, R is the number of districts and t is the number of time periods and N=RT

Panel Unit Root Test

The objective is to test whether a given series is non-stationary for all the individual units' districts in our case). It is assumed that the series follows a general panel data model structure (for i districts, and t periods)

$$y_{it} = (1-\phi_i)u_i + \phi_i y_{i,t-1} + \varepsilon_{it}$$

i =1...........N;

t = 1 ....T

Climate Models

Climate models are computer-based simulations that use mathematical formulas to re-create the chemical and physical processes that drive Earth's climate. To "run" a model, scientists divide the planet into a three dimensional grid, apply the basic equations, and evaluate the results. Atmospheric models calculate winds, heat transfer, radiation, relative humidity, and surface hydrology within each grid and evaluate interactions with neighboring points. Climate models use quantitative methods to simulate the interactions of the atmosphere, oceans, land surface, and ice.

General Circulation Models (GCMs)

General Circulation Models (GCMs) are a class of computer-driven models for weather forecasting, understanding climate and projecting climate change, where they are commonly called Global Climate Models. Three dimensional GCM's discretize the equations for fluid motion and energy transfer and integrate these forward in time. They also contain parameterizations for processes - such as convection - that occur on scales too small to be resolved directly. Atmospheric GCMs (AGCMs) model the atmosphere and impose sea surface temperatures. Coupled atmosphere-ocean GCMs (AOGCMs, e.g., HadCM3, EdGCM, GFDL CM2.X, ARPEGE-Climate) combine the two models.

Dynamical Downscaling

A Regional Climate Model is a tool to add small-scale detailed information of future climate change to the large-scale projections of a GCM. It is a comprehensive physical model, usually of the atmosphere and land surface, containing representations of the important processes in the climate system (e.g. clouds, radiation, rainfall, soil hydrology) as are found in a GCM.

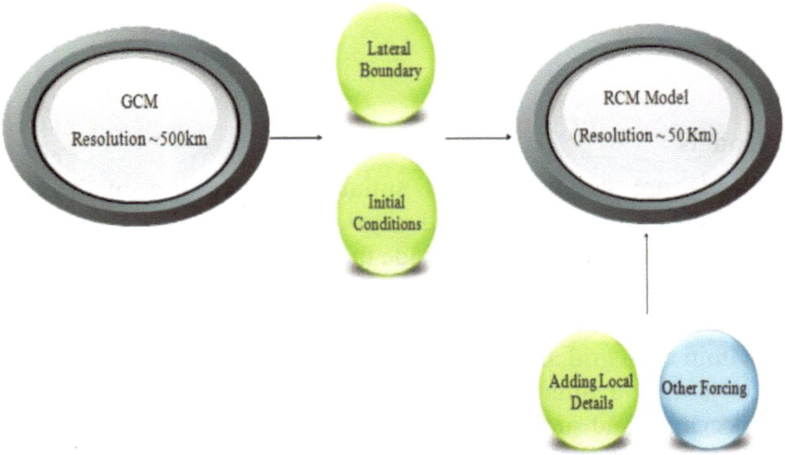

Climate Change Models on Marine Ecosystems

- Changes in Global Marine Primary Production
- Changes in Fisheries Production
- Changes in Fish Population Processes
- Changes in Fisheries on Human Societies
- Changes in Sea Level Rise
- Changes in Marine Ecosystem Productivity
- Climate change and its impacts on Aquaculture

Atmosphere-Ocean General Circulation Models (AOGCMs)

AOGCMs examine which aspects of the models determine how ocean biology responds to climate. (Sarmiento *et al.*, 2005; Gregg *et al.*, 2003; Richardson, and Schoeman, 2004). The study compared "realistic" emission scenarios for the period from pre-industrial to 2050 and 2090 with a control in which emissions remained at pre-industrial levels. Three groups of factors which govern the biological response are:

- Warming
- Light supply, as determined by ice cover, cloudiness and surface mixed-layer thickness
- Altered nutrient supply due to changed vertical stability and exchange

Earth System Model

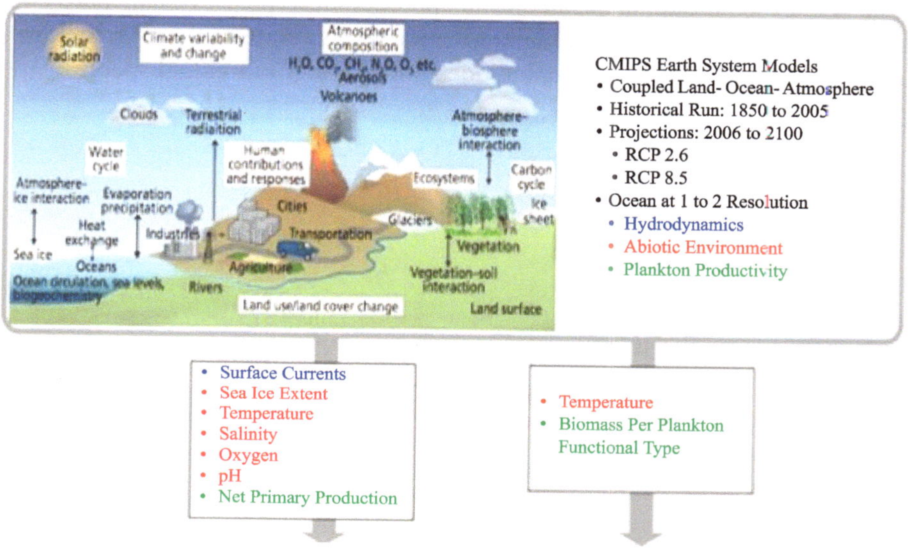

Dynamic Bio-climate Envelope Models (DBEMs)

DBEM is a mechanistic species distribution model that links prediction of habitat suitability to their spatial and temporal population dynamics and eco-physiology. Lam V.W.Y. *et al.* (2016) applied the DBEM to project changes in distributions, abundances and catches by the mid-21st century for each of the 887 marine fish and invertebrate species on a half degree x half degree grid of the world ocean (280 EEZs and the high seas) under the Representative Concentration Pathways (RCPs) 2.6 and 8.5 (Method).

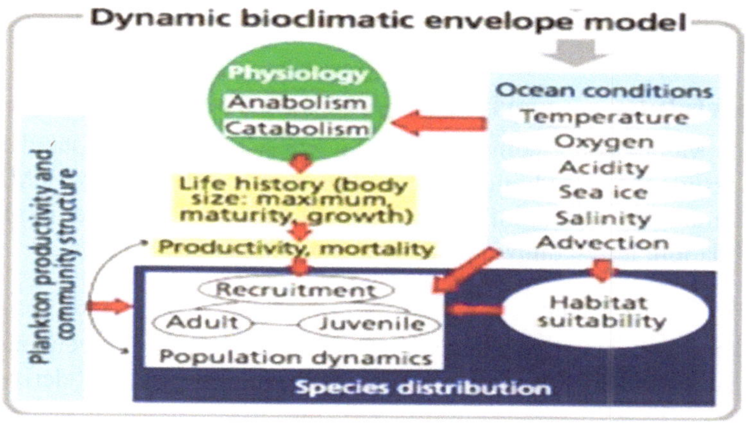

Dynamic Size-based Food Web Model

The size-based model is based on the approach described by Blanchard *et al.* (2012) and uses the same parameterization unless stated otherwise below. It distinguishes fish by their size, but does not consider individual species.

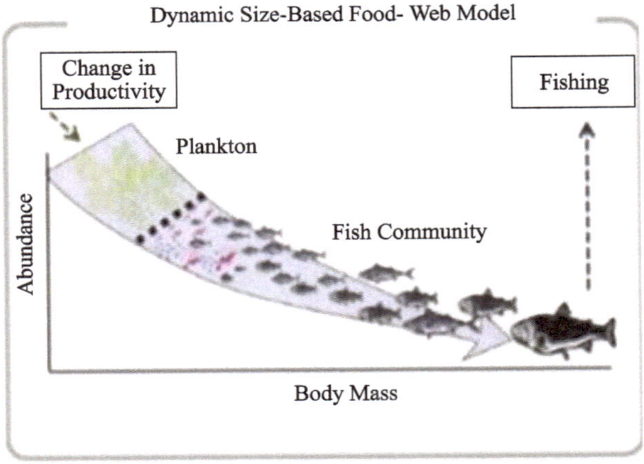

CORDEX Model Outputs (Freely downloadable)

The Coordinated Regional Downscaling Experiment (CORDEX) was initiated by the World Climate Research Programme (WCRP) in 2009 in response to the need for a coordinated framework for evaluating and improving regional climate downscaling (RCD) techniques and producing a new generation of RCD-based fine-scale climate projections for identified regions worldwide. The CORDEX framework additionally offered the potential for better coordination of RCD-related research and modeling activities within the regional climate modeling and downscaling communities and further to bridge the gap between the climate modeling community and end users of climate information across the globe.

Coupled Models Inter-comparison Project Phase 5 (CMIP5)

The Coupled Model Inter-comparison Project (CMIP) is a standard experimental protocol for studying the output of coupled atmosphere. CMIP provides a community-based infrastructure in support of climate model diagnosis, validation, inter-comparison, documentation and data access.

CMIP5 Model for Fisheries

CMIP5 based on Representative Concentration Pathway (RCP) of greenhouse gas emission scenario RCP4.5, RCP 6.5, RCP 8.5. Outputs of CMIP derived from

- The Geophysical Fluid Dynamics Laboratory Earth System Model 2 M (GFDLESM2M)
- The Institute Pierre Simon Laplace (IPSL) (IPSL-CM5-MR) and
- Max Planck Institute for Meteorology Earth System Model (MPI-ESM MR) (Method)

Adaptations

Step 1. Farm- Level Costs
and Earnings for 4 Production Systems
(2 Catfish, 2 Shrimp)

↓

Step 2. Expert Elicitation:
Change in last 10 Years, Perceptions for the Next
10 Years and Attribution to Climate Change Impacts

↓

Step 3. Direct Costs + Autonomous Adaptation Costs,
and Benefits, at farm Level for the Next 10 Years,
and Beyond to 2050

Adaptation Experiences from FAO and Partners

Reduced Yields :
- Secure Access to Higher Value Markets;
- Selective Breedings for Faster Growing Strains or for Disease Resilient Strains;
- Improve Water-Use Efficiency and Sharing (e.g. With Users of Irrigated Rice Paddy);
- Invest in Aquaculture Infrasture Improvement (e.g. Net Cages and Raised Dykes in Flood Prone Pond Systems).

Increased Yield Variability :
- Diversify Livelihoods (e.g. Ecotourism);
- Shift to Culture Based Fisheries or Shift to Hatchery Seed for Previously Wild Caught Seed Stocks.

Increased Risks :
- Insurance;
- Early Warning, Monitoring and Communication;
- Improved Safety at Sea.

Increased Vulnerability :
- Soft Defences (e.g. Wetland Rehabilitation);
- Disaster Risk Reduction and Responce.

Mitigations

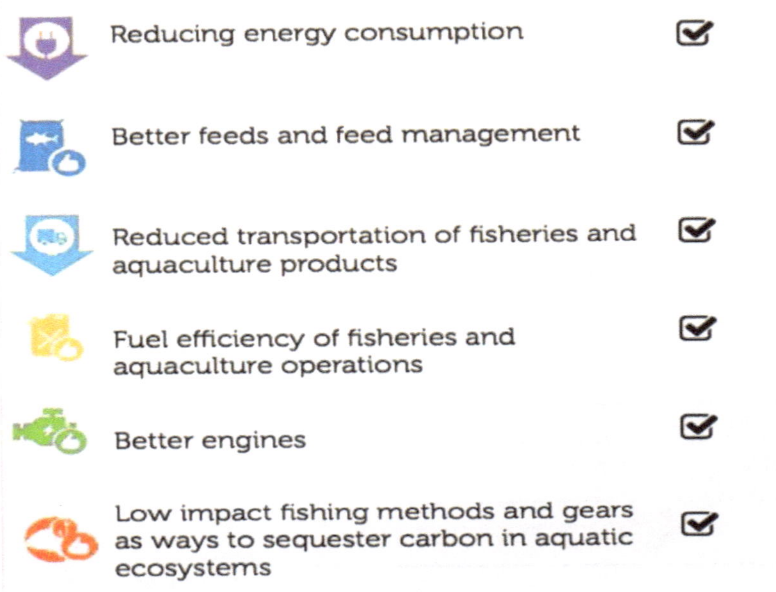

To conclude

To assess the impact of climate change, fit the regression function / Statistical function / Econometric function initially, then download the model outputs (whichever applicable) and finally plug the model outputs into regression function and estimate the impact.

20

Unique Approaches in Impact Evaluation

Sivaramane N.

Introduction

Impact Assessment is simply defined as the process of identifying the future consequences of a current or proposed action. The "impact" is the difference between what would happen with the action and what would happen without it (International Association for Impact Assessment). It's a structured process for considering the implications, for people and their environment, of proposed actions while there is still an opportunity to modify (or even, if appropriate, abandon) the proposals. It is applied at all levels of decision-making, from policies to specific projects.

IA aims to

- provide information for decision-making that analyses the biophysical, social, economic and institutional consequences of proposed actions;
- promote transparency and participation of the public in decision-making;
- identify procedures and methods for the follow-up (monitoring and mitigation of adverse consequences) in policy, planning and project cycles; and contribute to environmentally sound and sustainable development.

In its broadest sense, impact assessment is the process of identifying the anticipated or actual impacts of a development intervention, on those social, economic and environmental factors which the intervention is designed to affect or may inadvertently affect. It may take place before approval of an intervention (ex ante), after completion (ex post), or at any stage in between. Ex ante assessment forecasts potential impacts as part of the planning, design and approval of an intervention. Ex post assessment identifies actual impacts during and after implementation, to enable corrective action to be taken if necessary, and to provide information for improving the design of future interventions.

A distinction can be made between two separate but interlinked levels:

a) Internal monitoring and evaluation for ongoing learning, through for example the integration of specific impact indicators into existing management information systems, which makes information immediately available to staff;

b) External impact assessment, often involving independent investigators. Such assessments produce reports for specific purposes, such as poverty impact assessment, regulatory impact assessment, social impact assessment or health impact assessment. Certain types of ex ante assessment may be part of the approval process for certain types of intervention, including environmental impact assessment and economic impact assessment (cost-benefit analysis). These may contain their own ex post monitoring activities. Separate ex post assessments may be undertaken or commissioned for any particular intervention or set of interventions, to provide fuller information than may be available from routine monitoring and evaluation.

In the context of sustainable development, the social, economic and environmental impacts of an intervention are all interlinked. The various types of impact assessment may therefore need to be combined in an integrated impact assessment, whose nature will vary according to the type of intervention, and the aims and cost-effectiveness of the overall impact assessment package.

For each impact assessment type, a wide range of methodologies has been developed, according to the precise purpose of the assessment, the types of question to be asked, the organisational context, the socio-economic context, available budget, research capacity and other factors. An impact assessment may include any or all of:

a) Quantitative statistical methods involving baseline studies, the precise identification of baseline conditions, definition of objectives, target setting, rigorous performance evaluation and outcome measurement. Such methods can be costly, limited in the types of impacts which can be accurately measured, and may pose difficulties for inference of cause and effect. Some degree of quantification may be necessary in all impact assessments, in order to evaluate the success of the intervention and the magnitude of any adverse effects.

b) Qualitative methods suitable for investigating more complex and/or sensitive types of social impacts, e.g. intra-household processes, policy issues and investigation of reasons for statistical relationships and policy implications. These methods generally require high levels of skill, and may be relatively costly. Some degree of qualitative interpretation may be necessary in all impact assessments, in order to evaluate the causes of impacts which have been observed.

c) Participatory approaches suitable for initial definition or refinement of the actual or potential impacts which are of concern to stakeholders, questions to be asked, and appropriate frameworks and indicators to be used. Such approaches can contribute to all types of assessment, and are particularly suited to exploratory low budget assessments and initial investigation of possible reasons for observed statistical relationships. They offer a means of involving stakeholders in the research, learning and decision-making processes. These methodologies also require a certain level of skill, depending on the issues to be addressed and ways in which they are integrated with other methods. Some degree of stakeholder participation is likely to be necessary in all impact assessments, in order to achieve a good understanding of stakeholder perceptions of impacts.

Whatever mix of techniques is used, consideration should be given to:

- Transparency and public accountability
- Stakeholder involvement
- Reliability of the information obtained
- Reliability of inference for policy improvement
- Cost and skill requirements

Outputs, Outcomes and Impacts

It is important to distinguish the finer differences between the terms - output, outcome and impact so that monitoring, evaluation and impact assessment can be easily understood. The following are contributions of various authors in distinguishing the terms, albeit in different dimensions or using different terms, though they all mean the same.

Logic framework

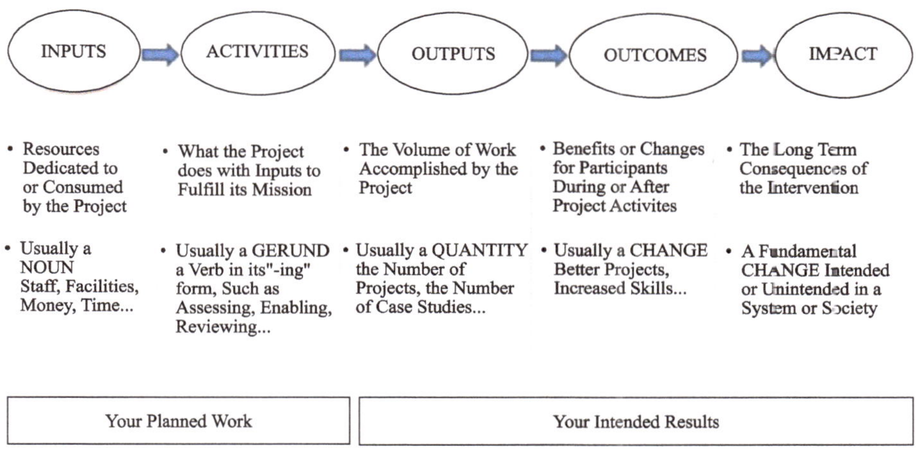

The Internal Logic of the Developmental Results Chain

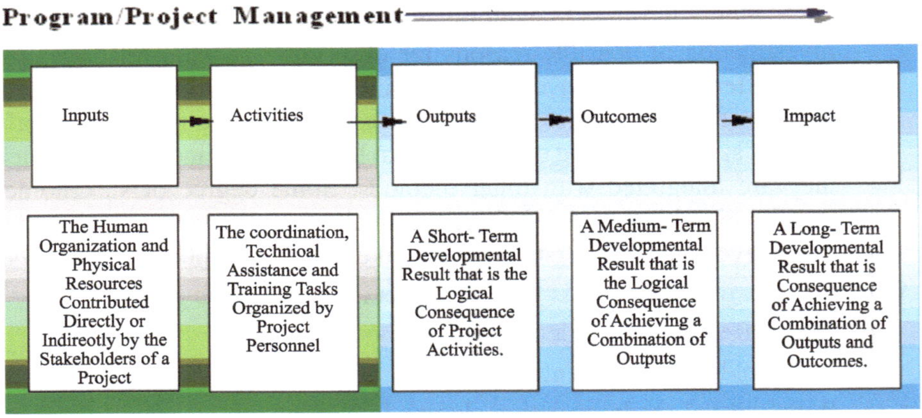

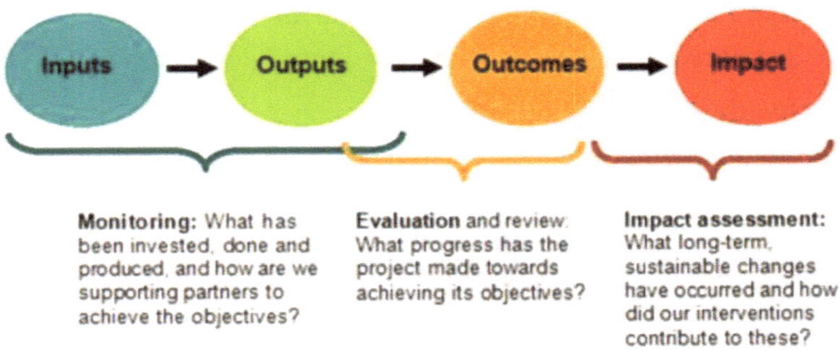

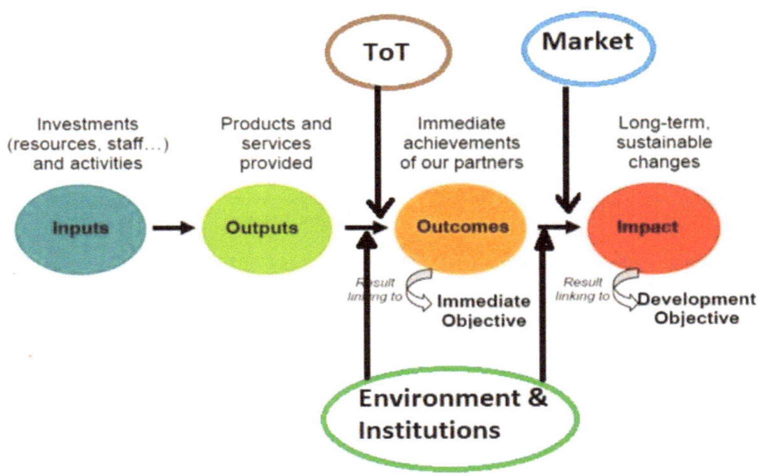

Much has been written about the socio-economic benefits and competitive advantage that developed countries achieved by investing in scientific research and technological innovation. For developing and emerging economies, it is recognized that sustainable development depends on establishing and supporting R&D institutions that both perform good science and share their knowledge and technology outputs.

A return on R&D investment, and the magnitude of that return, is important to policy-makers, tax payers, government officials, development agencies and, of course, those funding the research and the research institutions themselves. It should be clearly noted that the following notes does not evaluate the performance of Technology Transfer Offices, but rather focuses on determining the socioeconomic impact of transferred knowledge and technology.

R&D institutions in developing countries operate with limited financial resources for R&D and even less funding for technology and knowledge transfer. The socio-economic challenges experienced by developing countries put more pressure on R&D institutions, requiring them to effectively and efficiently address local, social and economic development needs through the transfer and adoption of innovative science. To this end, a key responsibility of research institutions in developing countries is to make research outputs available for use by society and local industry. It is therefore critical that research institutions not only generate relevant research, but also transfer and diffuse research results in a way that maximizes impact. A well-developed and comprehensive monitoring, evaluation, and impact assessment framework is necessary to measure efforts by institutions to meet R&D objectives. Such a framework can assist research institutions in:

- Improving the efficiency of research resource allocation
- Improving the standard and effectiveness of project decision-making
- Directing future research plans more effectively
- Obtaining evidence of resource mobilization
- Prioritizing research based on the level of economic returns and positive social impact

Technological innovation transforms an idea generated during research into a new or improved product that can be introduced into a market, a new or improved operational process used in industry and commerce, or a new approach to a social service. Monitoring, evaluation, and impact assessment should be conducted throughout the R&D continuum described below:

- **Research and technology generation**: Basic research, applied research, and experimental development are included.

- **Technology development**: During this stage, knowledge from research is combined with practical experience to direct the production of a new product.
- **Technology adaptation**: This entails piloting technology and simulating real-life conditions for the production of the technology are typically involved.
- **Technology transfer**: An important component of technology transfer is IP (intellectual property) management. Typically, institutions manage IP protection, routes to commercialization or transfer, and contractual arrangements that facilitate the transfer of intellectual property from the lab to the market.
- **Technology adoption and diffusion**: This stage of the process is key, for it signifies the point that products, transferred to the market, achieve depth and spread widely. Technology *adoption* is measured at one point in time and is associated with the use of transferred technology; technology *diffusion* is the spread of a technology across a population over time.

A robust monitoring, evaluation, and impact assessment framework should demonstrate transparency and confer accountability. It is therefore important that systems enable institutions to document, analyze, and report on research and technology transfer performance effectively.

Important tasks in Impact Evaluation

a) ***Developing framework:*** Blueprint of the framework

b) ***Identification of indicators:*** Crucial part in IA. If right indicators are identified, it will end up in doing the IA clearly and objectively and moreover, only few indicators will result in the final output as expected.

c) **Attribution**: Very tricky and infers how the outcome is attributed due to the particular intervention. There are some methods through which the effect of other variables could be culled out and trying to give attribution for which it has been intervened.

To develop framework, the following questions are to be asked:

- To whom? (Stakeholder)
- Why we? (Rationale why we are doing this)
- How long? (Duration of the stimulus)
- What category? (Training/ Research / Extension)
- What to assess (KAS/pub./tech.)? Up to which level?
- When to start IA (RPP?)? When to assess impact?
- How (bespoke)

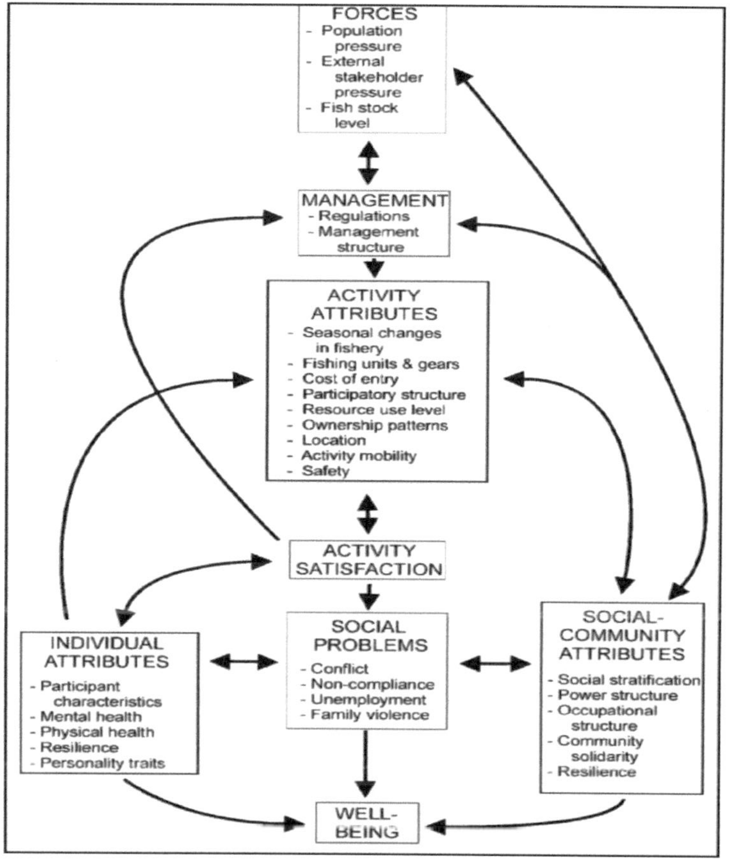

The ultimate measure is the well-being of the fishers/society/community

Model framework in marine fisheries (Socio-economic IA framework for Marine Fisheries)

Indicator

Indicators are the signs that show the extent of change that resulted from projects or programmes. Indicators help to measure what actually happened in terms of quantity, quality and timeliness against what was planned. Indicators should be SMART and are established at the formulation stage by stakeholders and programme managers to help them measure the extent to which the project interventions contribute to outputs and outcomes. Indicators should be selected in participatory mode with the consent of the stakeholders. Once identified, baseline study should be conducted.

SMART Indicators

- Specific (who, what, where, when & how)
- Measurable (counted / observed / analysed / tested or challenged)
- Attainable / achievable (straight-forward & cost effective)
- Relevant (linked to outcome)
- Trackable / time-bound (time-frame & frequency of data)
- Parsimony/objective-oriented/ satisfactory to client

Examples for indicators

a) Input indicators

- Number of mentors/coaches mobilized
- Number of hours mentors/coaches spend preparing and delivering activities

b) Process Indicators

- Availability of common methodology for mentors/coaches
- Activities to inform the target group of the existence of the programme

c) Output Indicators

- Number of young people who contacted the programme
- Number of young people who took part in one session

d) Result indicators

Quantitative

- Share of participants who defined their objectives during the programme
- Share of participants who reached at least one of the objectives during the programme

Qualitative

- Narratives from beneficiaries on what differences the coaching made to them
- Narratives from mentors or coaches on what differences the coaching made to the beneficiaries

e) Impact Indicators

Quantitative

- Share of participants who stay in education and training 3 months/6 months after the programme
- Share of participants who successfully complete their education / training

Qualitative

- Narratives of beneficiaries about how the programme helped them stay in education and training or complete their programme
- Narratives of mentors or coaches of how the programme helped beneficiaries stay in education and training or complete their programme

Impact Assessment Methods and Techniques

Impact type	**Method**	**Technique**
Intermediate impact • Institutional changes • Changes in the enabling environment	Survey, monitoring	Simple comparison/trend analysis
Direct product of research	Effectiveness analysis using logical framework	Simple comparison: target vs. actual
Economic impact (micro, macro, spillovers)	Econometric approach, surplus approach	Production function, total factor productivity, index number methods, and derivatives
Socio-cultural impact	Socioeconomic survey/ adoption survey	Comparison over time
Environmental impact	Environmental impact assessment	Various • Qualitative • Quantitative

Ex-Post (Before-After analysis)

Before and after' experiments provide near accurate measurement of the impact of a technology/treatment. *e.g.* Change in calorie intake / income change due to the implementation of MGNREGA programme

Problems faced in this analysis:

- Natural change not accounted
- Detailed baseline survey to be conducted with clear assessment of the impact of intervention before the start of the project
- Not suitable for long-term programme

Treatment vs Control

This is more applicable in cases where factors can be fully controlled and very less influence from extraneous factors. However, setting a true control is very difficult in social sciences. It is assumed in the model that the treatment varies from control only on the particular attribute for which treatment effect is studied.

E.g. The age, education level, income, etc. of sample in treatment should match with that of sample in control.

Particulars	**Treatment**	**Control**
Age	35	55
Education level	3	2
Irrigation	Yes	No
Yield	2000 kg/ha	1200 kg/ha

Is it fair to compare? If treatment and control differs, mean could not be compared directly because of the differences in attributes.

Difference-in-Difference method

It is a statistical technique used in econometrics and quantitative research in the social sciences that attempts to mimic an experimental research design using observational study data, by studying the differential effect of a treatment on a "treatment group" versus a control group in a natural experiment. It is a useful technique to use when randomization on the individual level is not possible.

Treatment		**Control**	
Before	**After 10 years**	**Before**	**After 10 years**
(A)	(B)	(C)	(D)
₹ 10,000	₹ 18,000	₹ 10,000	₹ 16,000

Is there any significant change due to the intervention?

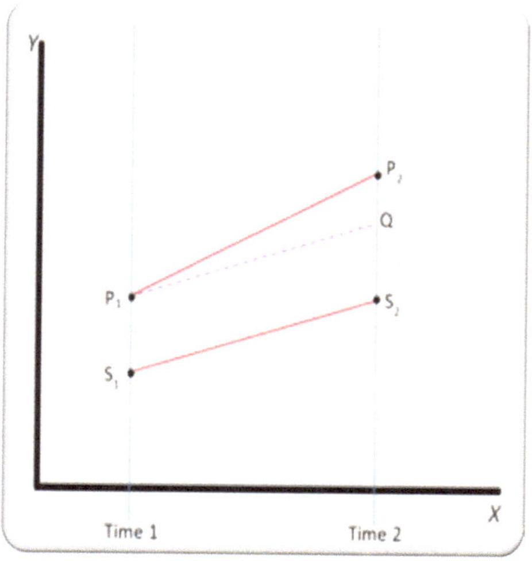

Randomized Controlled Trials (RCT)

A study design that randomly assigns participants into an experimental group or a control group. As the study is conducted, the only expected difference between the control and experimental groups in a randomized controlled trial (RCT) is the outcome variable being studied.

Advantages

- Good randomization will "wash out" any population bias
- Easier to blind/mask than observational studies
- Results can be analyzed with well-known statistical tools
- Populations of participating individuals are clearly identified

Disadvantages

- Expensive in terms of time and money
- Volunteer biases: the population that participates may not be representative of the whole
- Loss to follow-up attributed to treatment

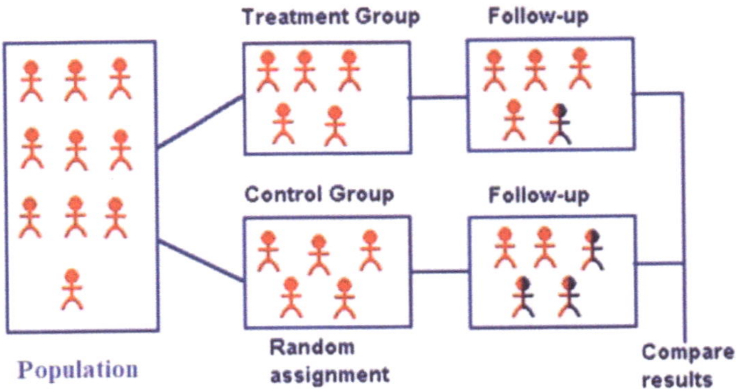

Propensity Score Matching (PSM)

PSM constructs a statistical comparison group that is based on a model of the probability of participating in the treatment, using observed characteristics. Participants are then matched on the basis of this probability, or *propensity score*, to non-participants. The average effect of the treatment (*ATT*) of the program is then calculated as the mean difference in outcomes across these two groups.

Necessary conditions

- Conditional independence (namely, that unobserved factors do not affect participation)
- Sizable common support or overlap in propensity scores across the participant and nonparticipant samples.

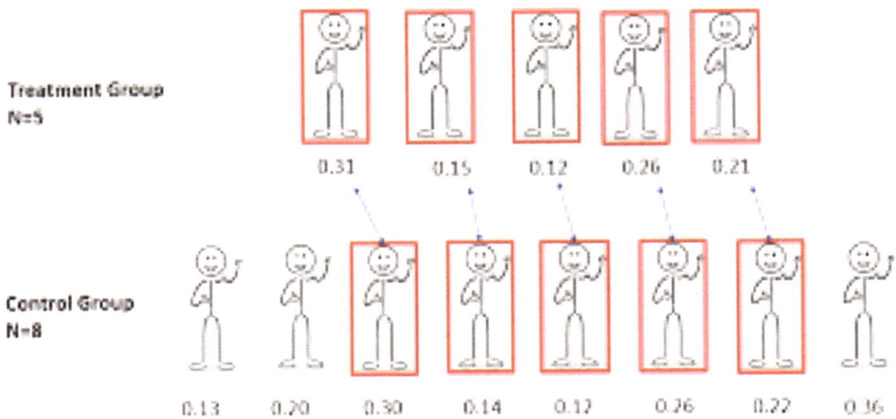

Common support in PSM

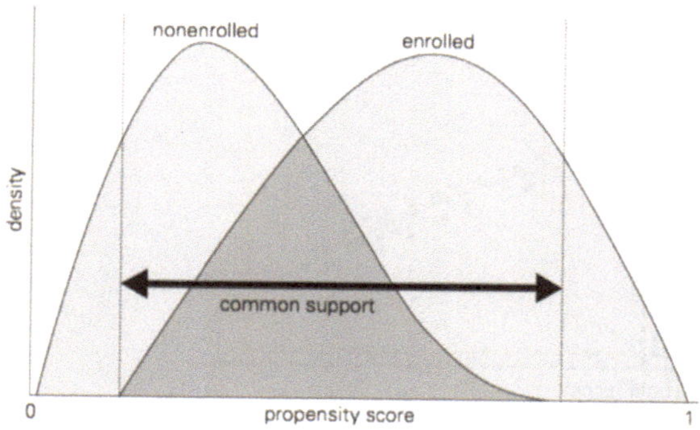

Matching methods

Kernel method

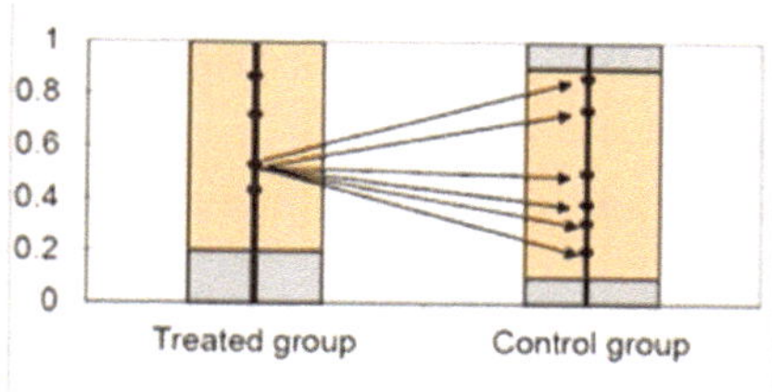

Nearest neighbor

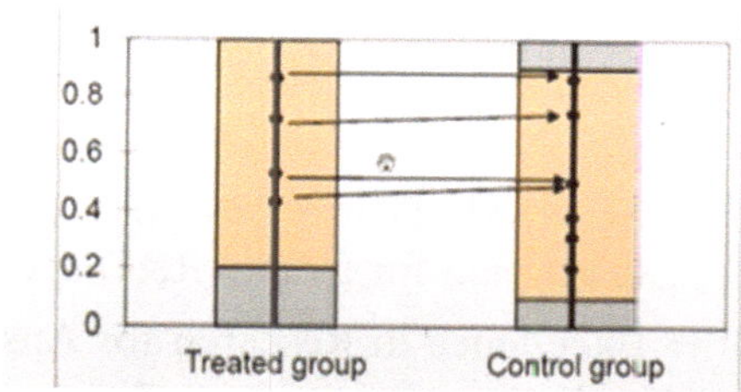

Regression Discontinuity

The samples across which to compare would have to be sufficiently close to the eligibility cut-off to ensure comparability. Furthermore, unobserved heterogeneity may be a factor if people within the eligible targeting range exhibit variation in actual take-up of the program, leading to selection bias. In that case, eligible and non-eligible samples close to the eligibility cutoff would be taken to compare the average program effect.

Examples

- Grameen Bank's microcredit is targeted to households with landholdings of less than half an acre;
- Pension programs are targeted to populations above a certain age

Particulars	Control	Treatment
Average off-take of credit per capita	□ 1,00,000	□ 10,000
Average calorie intake of children Kcal	1300	1100

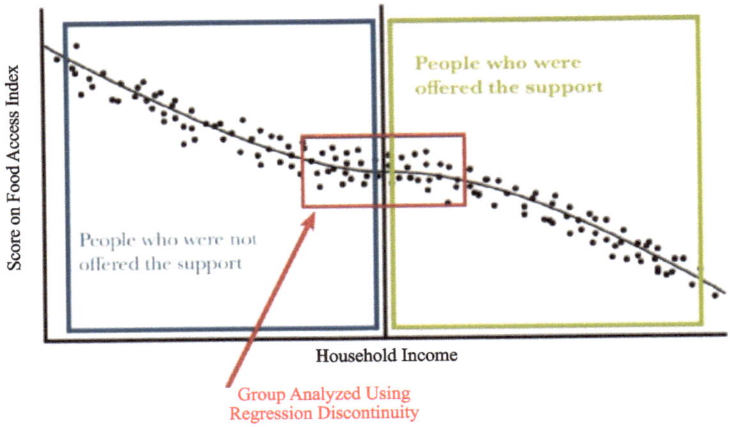

Frameworks

- Research Excellence Framework (REF), UK – assesses performance of UK universities to determine funding allocation
- STAR METRICS, US – uses data mining and other low burden methods to account for federal R&D spending
- Excellence in Research for Australia (ERA), AU – uses bibliometrics, and other quantitative indicators, to map R&D output
- Canadian Academy of Health Science (CAHS), CA – aims to provide consistency and comparability while retaining flexibility
- National Institute of Health Research (NIHR) Dashboard, UK – provides performance management information at various levels of aggregation
- Productive Interactions, EU – flexible approach to help institutions learn and improve their performance against their own goals

Impact measures

Production and economic impact measures

- Risk reduction
- Yield increase
- Cost reduction
- Reduction in necessary inputs
- Employment creation
- Implication for other sectors of the economy

Institutional impact measures

- Changes in organizational structure
- Change in the number of scientists
- Change in the composition of the research team
- Multidisciplinary approaches and improvements
- Changes in funding allocated to the program
- Changes in public and private sector participation
- New techniques or methods

Socio-cultural impact measures

- Food security
- Poverty reduction
- Status of women
- Increases in knowledge and skill level
- Number and types of jobs
- Distribution of benefits across gender and geographical locations
- Changes in resource allocation
- Changes in cash requirement
- Changes in labour distribution nutritional implications

References

https://sites.google.com/site/econometricsacademy/econometrics-models/propensity-score-matching #:~:text= Propensity %20score %20matching %20 is %20 used, targeted%20 at% 20 helping % 2 0particular % 20 schools.

Madeleine Hall-Arber et al. (2009) Figuring Out the Human Dimensions of Fisheries: Illuminating Models. Marine and Coastal Fisheries Dynamics Management and Ecosystem Science Management (1): 300-314. DOI: 10.1577/C09-006.https://the fishsite.com/articles/a-practical-guide-to-using-ai-in-aquaculture.

NILERD (2015), KVKs impact on Dissemination of Improved Practices and Technologies, Published by National Institute of Labour Economics Research and Development, New Delhi.

21

Methodological Issues in Fisheries Data Collection

Ramanujam Jayaraman

Introduction

Economists consider a producer as an income or profit maximiser. In our context, it may mean enhancement of fisheries contribution to the GDP, increased fish production, Nutritional security of those who need it most, not just those who can afford to pay for it improvement in the livelihood conditions of the stakeholders, augmenting exports to net more foreign exchange, growth with social equity or inclusive growth, etc. Realisation of a goal or target depends on development and enforcement of necessary policy strategies for which a reliable data base is inevitable. Analytical trends of the past data could help to project future trend which could indicate feasibility of the envisioned goal. Here, an outline of some of the Methodological issues in Fisheries data collection is provided.

The Resource Endowments

Fisheries is a sunrise sector in India, in terms of food and nutritional security, employment and income generation. India is the 3rd largest fish producer and the 2nd in aquaculture production, globally. It contributes about 1% of national GDP and 5 % of agriculture GDP. The exports of marine products account for about 5% of total Indian exports in 2018-19. The estimated fish production during 2017-18 was 12-59 million tones (mt). It was 10.14% increase over 11.43 million t in 2016-17. Inland fisheries grew at 14.05% while marine fisheries rose at 1.73% in 2017-18. Fisheries sub-sector provides livelihood support and gainful employment to 14 million people. Many more may be employed now as available data is yet to be updated.

Fisheries is the primary source of livelihood in coastal villages along the 8,118 km long coastlines of the country. Exports of marine products remain the backbone of fisheries contribution to the national GDP. Fisheries development is mostly export driven. Exports of Indian marine products have been registering a continuous, positive growth trend.

India is among the top five fish exporting countries in the world at third place preluded by China and Indonesia. About 1.1% contribution is of Fisheries sector in Indian economy and about 6.72% contribution is in Agriculture. In 2021-22, the country exported 1.36 million MT od seafood worth US $ 7.76 billion which is all time high export by value. India has exported marine products like frozen shrimps, fish, cuttlefish, squid and dried items, live and chilled items to 123 countries. Out of these, frozen shrimp is the major exported marine product contributing to more than 53.18% to the total quantity and about 75.11% of the total export value.

Essentially fisheries development in India has almost been shrimp exports driven fisheries development. Marine capture shrimps leveled off in mid-1990s. Hence, scope for enhancing marine landings significantly is restricted. India is a pioneer in shrimp farming and has a long history of shrimp farming.

Farm area under shrimp farming rose from 1.4 lakh ha in 2016-17 to 1.67 lakh hain 2020-21. Farmed shrimp production went up from 4.97 lakh tons in 2016-17 to 8.44 lakh tons in 2020-21. Tiger shrimp farm area decreased slightly from 58,851 ha in 2017-18 to 58,196 ha in 2020-21 whereas production also decreased

from 58,163 tons in 2016-17 to 27,616 tons in 2020-21. Scampi production was 9,983 tons in 2017-18. with a decrease of 8,303 tons in 2020-21. *Penaeus vannamei* shrimp is the major, exotic species farmed in the country. Vannamei shrimp farming over shadows Indian traditional shrimp farming. Vannamei farmed shrimp production increased from 87,252 ha in 2016-17 to 93,496 ha. in 2020-21. In quantity terms, it was 6,22,327 t in 2017-18 and increased to 8,15,745 t in 2020-21.

Methodological Issues – Inland Fish Production Estimates

Do we still need appropriate methodology for data collection?

Marine fish landing data sets published by the Central Marine Fisheries Research Institute are available for use for a fee but not all data are made available. The CMFRI follows multi-stage randomized block design to estimate the marine fish landings in the country. But, there is no mechanism available to cross check the data at village, panchayat, block, district or state level. Earlier, States used to collect their own data and will reconcile it with that of CMFRI and both will then agree on a data set for that state. States no longer do data collection on the marine fish landings in their States and the staff who attended to that work now look after welfare schemes.

There is couple of issues too. I am not competent to air them as an economist, but still worth to ponder over. Fish production estimates are generally with reference to stocks of fish population in an area usually called fishing grounds, in tamil it is called as "pars"…. Now landings data do not correspond with the fishing ground, that is, from where the catch originated? This is important from stock identification point of view as well as from FAO code of responsible fishing. Now importers can insist imports originate from sustainably managed fisheries.

Inland fish production records positive, steady growth trend. Inland fish production accounts for about two third of total fish production. Reservoirs have a vast area, yet reservoir fish productivity is low as the reservoirs are not managed scientifically at all. The State fisheries department stocks fish and let the co-operative society fishermen harvest it and yield is shared perhaps in 60 -40 ratio. The erstwhile CIFRI came with a methodology for estimation of fish production from reservoirs. Big reservoirs are leased out to big farmers or just businessmen who underreport the actual yield netted. Population dynamics of reservoirs have changed a lot in the last few decades. Tilapia is rampant now in many, if not most, reservoirs. No staff or mechanism exists to estimate reservoir fish production correctly. With no planned scientific management reservoirs are not managed and data available for fish production from them is not reliable.

We have very long rivers, some of them perennial. Rivers have turned to be the reservoirs of sewerage, agricultural run-off laden with huge quantum of pesticides and insecticides, industrial effluents discharge. So riverine fisheries is under huge stress and whatever data we have on fish production from rivers is unreliable and no cross check is made. River reclamation programmes were not there and we lost riverine bio-diversity significantly.

Now we need to turn to the aquaculture sector which is the game changer in fish production. India is a carp country. Carps are reared in ponds and farms in a variety of geographical locations. We have no data on how many freshwater farms are there, what is their total area and the actual area under aquaculture, the actual quantum of fish produced from the ponds at pancahyat or district or state level. Total fish production data are reported but how it was collected and what is its accuracy remains largely questionable. Many water bodies are now bus-stands, houses, urban conglomerates, etc. No one knows what was there, what was lost and what remains –in terms of freshwater bodies. So, we have no reliable data on the extent of freshwater area available currently for aquaculture of freshwater fishes without which how could we estimate the farmed fish production. The Central Inland Capture Fisheries Research Institute (CICFRI), Barrackpore has been working on this using satellite data. Anna University Institute of Remote Sensing also worked on this. The information that we could make use of is not just available. Like CMFRI, they may charge a fee for sharing data in their own way.

Farmed shrimps are pooled together by the buyers and it is very much difficult to pinpoint how much originated from each destination, that is farm-wise or village-wise production details. The buyers may have the data but they will not share for obvious reasons. That is one reason why we are unable to track the source of antibiotic residue at specific farm level. The MPEDA collects farm samples regularly as per the European market needs. But, ground level mechanism is absent. Neither state level nor national level data with split-up details at panchayat or district level are available.

Next is Our Shrimp Farming Sector

The Coastal Aquaculture Authority (CAA) is the national regulator of coastal aquaculture. It is also the registering body. The CAA is making available data in public domain on the farms registered, area reportedly under farming for these registered farms. But, as you all know more than me, the actual area under farming is not known. MPEDA is also registering aqua farms but that data is also not the full picture. MPEDA does some back calculation on the farmed shrimps quantity based on the shrimp exports data which it has.

The last case we take is export data. The MPEDA is the designated regulator of marine products exports. It collects and makes available export data in public domain, item-wise, marketwise and port-wise. We do not know the origin of marine products exported from a particular port or exported to a particular market. The data available for export earnings from a State may refer to that netted through the ports or airports in that particular state but not necessarily produced by it.

Methodology and analytical techniques alone are insufficient when the data set lacks accuracy. We embarked on planned development of our economy. That is we charted our development through five year plans initially now with annual plans. Plans specify goals, strategies and policy guidelines and support. Without reliable data base how could the plan be realistic?

So we need to strengthen our data collection mechanism at all levels which is urgent and most important now than ever before.

22

Writing Research Papers: Breaking Myths and Building Right Perspectives

Krishnan P.

Introduction

Precise learning of the components of the general structure of a research article is inevitable in writing a research article. The general structure of a research article consists of Title, Abstract, Keywords, Introduction, Methods, Results and Discussion, Conclusion, Acknowledgements, References and Supporting materials. There are no specific guidelines for which one of the above should be written first.

Process of Writing - Building the Article

To start writing a research article, it is not advisable to consider all the tables and data that are available to the writer. The easiest way to start with is to choose the tables and data that can be used wisely in the paper. The writer should be aware of the types of the chart or table which are accepted and the number of charts and tables that should be included in the paper (generally a research paper contain 7 to 9 figures and charts in each category). It is also not recommended to use conventional ways of presenting data and figures (like bar charts, pie diagram). The figures or illustrations should not be so simple. Figures and Tables should be self-communicating. The reader of the paper is very well educated and may have mastery in the particular subject. The language and illustration should be appealing to them. The figures and illustrations should have an impact on the reader to think and draw inference from the data provided by the research. Two dimensional and three dimensional data presentation create more interest in the reader rather than the one dimensional presentations. Next to consider is ***Methods*** that explain how the data was collected. In case of writing the methods in research paper, unlike thesis, the methods need not be elaborated in the paper. Results are the description of the data and the broad observation made. Results should not contain the inferences that are already known. Results should be supplementing the tables and charts, not a description of them. Discussion involves the comparison of the obtained result and the previous results. The contribution to the previous research results can be discussed. New findings and solutions can also be discussed. Conclusion is a single paragraph which altogether summarizes the final outcome. Generally the conclusion should not contain citations.

Introduction

Introduction is not recommended to be written first. It should be written after understanding the whole content of the paper.

Title, Abstract and Keywords

When these are to be written, the person who writes has the whole information and it is the place where the editor goes first in search of mistakes.

Are social Scientists Disadvantaged in the Impact Factor (IF) Race?

Some disciplines in the subject may not provide an impact factor for publishing as journal. The impact factor is a function of the citations occurring in the approved paper. More the number of research papers in a particular subject more the number of citations. Less the number of research papers in a particular subject less the number of citations (e.g. Mathematics). It is also a wrong perception that the person who is on other streams of research cannot publish in a high impact factor. For example, the impact factor of SCIENTIFIC DATA is close to 6. Most of the people very often will not decide that the paper written in the above topic would qualify to be published in the journal with an impact factor of 6. This myth in the mind of people should be eradicated.

SCIENTIFIC DATA

OPEN **Data Descriptor: Survey-based socio-economic data from slums in Bangalore, India**

Debraj Roy[1,2], Bharath Palavalli[3], Niveditha Menon[4], Robin King[5,6], Karin Pfeffer[3], Michael Lees[1,7] & Peter M.A. Sloot[1,2,7]

Received: 19 June 2017
Accepted: 28 November 2017
Published: 9 January 2018

In 2010, an estimated 860 million people were living in slums worldwide, with around 60 million added to the slum population between 2000 and 2010. In 2011, 200 million people in urban Indian households were considered to live in slums. In order to address and create slum development programmes and poverty alleviation methods, it is necessary to understand the needs of these communities. Therefore, we require data with high granularity in the Indian context. Unfortunately, there is a paucity of highly granular data at the level of individual slums. We collected the data presented in this paper in partnership with the slum dwellers in order to overcome the challenges such as validity and efficacy of self reported data. Our survey of Bangalore covered 36 slums across the city. The slums were chosen based on stratification criteria, which included geographical location of the slum, whether the slum was resettled or rehabilitated, notification status of the slum, the size of the slum and the religious profile. This paper describes the relational model of the slum dataset, the variables in the dataset, the variables constructed for analysis and the issues identified with the dataset. The data collected includes around 267,894 data points spread over 242 questions for 1,107 households. The dataset can facilitate interdisciplinary research on spatial and temporal dynamics of urban poverty and well-being in the context of rapid urbanization of cities in developing countries.

The **2018** journal metrics for **Scientific Data** are: 2-year **impact factor**: 5.929. 5-year **impact factor**: 6.776. Immediacy index: 1.014.

Who Can Write a Review Paper?

Generally, the convention is that the review papers are written only by experts who have a great expertise in a particular subject. But the truth is that about 80% of the journals of the major publishing houses accept reviews from anybody. Publishing of good research paper requires the best homework of analyzing all the aspects of the paper and coming out with new inferences. Review paper is the starting step to the world of research papers.

What is the Area of Specialization of a Researcher?

The Ph.D. holders don't say that they are specialists in a particular subject (say, taxonomy) instead they say they are specialists in the particular section of that subject (crustacean taxonomy) in which they have been working independently. This understanding is not seen in the field of agriculture due to the fact that there are a number of specializations in masters and Ph.D. in agriculture. Incorrectly; it is considered that the person is a fishery biology specialist or an aquaculture specialist. It is better to concentrate on a particular division for which the researcher has his own interest and in which he could claim he is an expert. The area of specialization should not be confused with Masters' specialization.

Authorship Dilemmas

Author is somebody who have completed the research and contributed significantly for outcome of the research and who has a specific role in approving the paper. First author is someone who have completed and implemented the whole study. There is a common misconception that the corresponding author is someone who correspond the journal. Corresponding author is the person who is competent to answer which are posted to the authors after the paper is published. In the academic research, the Ph.D. guide will be the corresponding author. Generally, the last author is the corresponding author. In case of research done by institutions such as ICAR, the first author may be the corresponding author. Though corresponding author is a person who is to answer the questions after the publication, the person who submits the paper in the online portal for submission is called the corresponding author. CRediT (Contributors role taxonomy) is high level taxonomy including 14 roles that can be used to represent the roles typically played by contributors to scientific research. The role describes each contributor's specific contribution to the scholarly output.

How to Choose a Right Journal? How much the Impact Factor Matters?

The success of the paper depends on the journal metrics viz., Cite score, Impact factor, 5-year impact factor, Source Normalized Impact Factor (SNIP) and Scimago Journal Rank. A journal with a high impact factor may carry a paper which may not be in cited ones and on the other hand, the journals with very low impact factor which may be having citations more than 100 or 300 in a short period. In many publications, there are many papers which go uncited even once. This understanding is very important for choosing a journal based on impact factor.

$$\text{IMPACT FACTOR} = \frac{\text{Citations on the last two years' paper in the current year}}{\text{Total number of papers published in the last two years}}$$

5-year impact factor is nothing but which considers the papers of last 5 years instead of 2 years unlike impact factor.

$$\text{5 YEAR IMPACT FACTOR} = \frac{\text{Citations on the last five years' paper in the current year}}{\text{Total number of papers published in the last five years}}$$

The same goes for cite score which considers three years.

Journals of different specializations should be compared within the subject only. It is not advisable to compare between different subjects as both are different in their nature (using impact factor).

Source Normalized Impact per Paper (SNIP) is an effective way of comparing journals of different subjects. In the method of determination of SNIP, the citation patterns of the journals are grouped according to their subject.

$$\text{Source Normalized Impact per paper} = \frac{\text{Journals citation count per paper}}{\text{Citation potential in its subject field}}$$

Then each of them is normalized with the highest citation among them. Finally the journal is compared with mean of each group. This means the journals can be compared across the subjects using SNIP.

Scimago Journal Rank not only takes into account the number of citations made on a paper but also the prestige of the paper. For example, if a paper is cited by an individual of economics and also by an individual of natural science, the weightage given for the citation from natural sciences.

Who Downloads the Articles?

The best journal is the one which would help our articles to reach the right clients who are likely to cite the paper. It is not advisable to reject a journal with a bad impact factor. Blindly no one should depend on the impact factor. Each publishing house has its own online tools like **Journal finder**, **Journal suggester.**

Always try to improve the author metrics rather than journal metrics

Important author metrics include:

- H-index
- i10-index
- Research Gate score
- Citations

The author metrics do not decrease; either it remains still or increasing.

Ethics of Scientific Writing

Major unethical practices in scientific writing are falsification, fabrication and plagiarism. Falsification occurs when the author, though the data collected is not supporting the hypothesis, tries to stick to his/her hypothesis. Author does this by changing the data in accordance to the conclusion. Fabrication is the complete alteration of the data or the data and results are completely made up by the guesswork of the author. It is also known as 'cooking of data'.

Plagiarism may occur unknowingly in your paper which could have a drastic effect in your later life.

e.g.

Author should be aware of plagiarism. Text based similarities could be overcome with iThenticate and Grammerly softwares. The fields ***to be cited*** include translation peppered with own imagination, idea re-written in fully own words and inspired by free information in web (Google). Widely known/ common knowledge / proverb and our observations and inferences from our work / experience are the fields not to be cited.

Index